ENVIRONMENTAL SYSTEMS
AND ENGINEERING

by

JUDITH BOWER CARBERRY

Department of Civil Engineering

University of Delaware, Newark DE 19716

SAUNDERS COLLEGE PUBLISHING
Philadelphia   Fort Worth   Chicago   San Francisco
Toronto   Montreal   London   Sydney   Tokyo

Acquisitions Editor: Bob Argentieri
Managing Editor: Carol Field
Manager of Art and Design: Carol Bleistine
Art Director: Christine Schueler
Art and Design Coordinator: Doris Bruey
Cover Designer: Larry Didona
Director of EDP: Tim Frelick
Production Manager: Charlene Squibb

Cover Credit: WIDE WORLD PHOTOS

Printed in the United States of America

Environmental and Systems Engineering

0-03-29657-9

0123 090987654321

This work is dedicated to
my elderly parents
Rose and Charles McClintock
whose faithful support and
encouragement permitted its completion.

PREFACE

This book has served as class notes for the required Civil Engineering junior level one semester course in Environmental Engineering. Previous texts required for this course have been either too advanced or too elementary. In either case, they contained far too much material to cover in one semester, and it was necessary to skip vast portions of these texts.

When this book was used as class notes, the students were told that they were responsible for all the material contained therein. They read every word. They made suggestions and corrected mistakes. They asked questions which indicated that the presumed great leaps of faith needed filling out. Ensuing drafts presumably corrected these problems, and the number of corrections and suggestions has decreased exponentially. The final draft was edited and printed using the TeX scientific software developed by Donald E. Knuth.

The material proceeds from very general considerations to specific topics utilized in environmental engineering. The prerequisites for this required introductory course here include two terms each of chemistry and physics and four terms of math.

Furthermore, the same required Environmental Engineering course is intended to provide basic information for those students who will not continue on in Environmental Engineering, as well as to provide the specific technical information for those students intending to continue on in the field. This course also is intended to help students decide if they want to specialize in this area.

Whatever area of Civil Engineering students choose, they need to become knowledgeable about the environment in order to function as an engineer. They will design, construct, and utilize systems which interact with the environment in any area of Civil Engineering. Since the early seventies, proposals for new construction projects must contain an environmental impact statement. This regulation has been imposed so that 'progress' does not destroy the environment or require, in the end, huge sums of money for remedial environmental restoration. Engineers must become cognizant of any remedial costs related to environmental impacts.

And, finally, this text concentrates on the environmental problems of man's interaction with water and its effects on the environment. Water is a unique resource, since we need a rather pure source for our use, and then we must return it to the environment in a condition which will not cause a severe detrimental impact. Most other resources need not be so pure for our use or are merely consumed and not returned to the environment. In addition, water is an important consideration for other civil engineering disciplines such as soils, oceans, hydraulics and structural engineering. In many cases, this material may be a useful reference when the students enter practice—environmental, or otherwise.

The author is truly indebted to her students for their inciteful contributions to this text. She is further indebted to the following colleagues: Dr. Peter J. Warter, Chairman of the Department of Electrical Engineering, who generously provided access to that department's advanced computer system when the rest of the University computing facilities were less than adequate; to Dr. Daniel J. Grim, the hardware and software computer genius for the entire College of Engineering, whose patience is matched by his sense of humor; to Dr. Jack P. Seltzer of the University Academic Computing Services, another genius who contributed both microscale programming solutions and macroscale approaches to engineering applications; and to Ms. Judy Joos who tirelessly drafted and re-drafted the figures.

TABLE OF CONTENTS

FUNDAMENTALS

Chapter 1

ENVIRONMENTAL SYSTEMS

A. Ecological Considerations

The natural biosphere contains land, air, and water masses. Each physical phase is interrelated to the other, so that an examination of one phase alone can not be considered as an exhaustive investigation. Each physical phase receives physical, biological, and chemical inputs to which both its living and non-living forms must respond. The response to these inputs is in the form of either an absorption or an expenditure of energy. Each physical phase exists in a dynamic state, always receiving inputs and giving off outputs to the other physical phases of the ecosystem. If the dynamic forces upon a particular phase are in balance, then a steady-state condition exists and there is no perturbation of the system. A **steady-state** condition is not the same as an **equilibrium** condition. In any one phase, a steady-state exists when the rate of production of some material equals the rate of destruction by a separate reaction. The steady-state condition can be illustrated by Equation 1.1.

$$A \rightarrow B \rightarrow C \qquad (1.1)$$

Often, in natural systems, these spontaneous reactions occur in a series of consecutive reactions, with the product of one reaction becoming the reactant in the next reaction. If the reaction rates of these reactions were equal, then, the concentration of the intermediate product (illustrated here as B) would not change and the system would be at steady-state with respect to that intermediate product. This is often the case with a small, constant discharge to the environment.

In contrast, the equilibrium condition is shown in Equation 1.2, as follows:

$$A \longleftrightarrow B \tag{1.2}$$

If the rate of the forward reaction equals the rate of the reverse reaction, no reaction would be observed and the system would be at equilibrium. This condition might exist if no discharges were made to the system, but such a case is rare in the natural environment.

More likely is the previous case, where low level pollution discharges are made to the environment, and the best we can hope for is a steady-state condition in which no undesirable intermediate product builds up to an intolerable level. If, however, an excessive pollution load or a continuously-increasing load is exerted upon any system, the resulting perturbation of this previously-balanced system will cause an 'abnormal' response, and the steady-state system will no longer exist. Examples of such undesirable overloading of a system are the discharge of wastewater (sewage) to recreational surface waters or discharge of excessive concentrations of fertilizing nutrients to freshwater lakes and streams. These two cases will be discussed in detail in Section C of this chapter.

Each phase of the biosphere has an inherent absorptive capacity for handling pollution discharges. The absorptive capacity is limited by the system's capability to maintain a dynamic steady state by the expenditure or absorption of energy in response to these pollution discharges. The engineer must learn, therefore, how to characterize the steady-state condition of the biospheric phase, how to predict an 'abnormal' response to a perturbating discharge, and how to treat a discharge in order to reduce the perturbation to the receiving phase in the ecosystem.

Here we will consider only the liquid phase of the biosphere. In the case of water and the impact of its discharge to the environment, an accurate determination of each of the following considerations is important:

1. Specification of the environmental quality desired and determination of fundamental biological, chemical and biogeochemical interaction controlling the desired quality.

3

2. Identification of the potential pollutants in the water and determination of how they would affect the desired environmental quality if discharged.

3. Evaluation of the current and future technology of wastewater treatment for the relevant pollutants.

It is difficult, however, to get much more than an instantaneous picture of a system, as each component can change both in time and geographical location. Modern computers have provided significant aid in simulating and modeling these systems.

Figure 1 illustrates a flow diagram for a system involving water removal, use, and subsequent discharge back to the environment. This diagram can be used to determine the impact of such a discharge to the environment. The inter-relationships of each component in the system must be identified. The dynamic nature of the system is apparent when one considers that at any given time the purified water quality may vary, consumer uses are changing, improved wastewater treatment technology is being developed and implemented, and the desired level of environmental quality is changing. The overall tendency of this system, however, is to reach steady-state in which the water quality supplied, its uses, and the applied wastewater treatment for any one instant are maintained in balance with the instantaneous environmental quality desired.

As indicated in Figure 1, the final decision for the acceptable or tolerable level of environmental quality will be established by the citizens themselves in collaboration with appropriate regulatory agencies. Typical considerations would involve the amount of financial obligation required to obtain or maintain an environment which is safe from a public health standpoint and is relatively satisfactory for public use and enjoyment. A steeper cost/benefit ratio would probably not be acceptable. Since it is prohibitively expensive to purify wastewaters to a zero discharge level of quality, the citizens must expect to discharge some acceptable level of pollution. In some cases, the quality of the treated wastewater discharge enhances the receiving water quality and in other cases the citizens must accept a somewhat reduced level of environmental quality.

Historically, the 1972 Clean Water Act stipulated that all major point discharges must have an acceptable level of water quality. All municipalities are required to carry out a stringent wastewater treatment program, and federal funds are provided to implement the construction

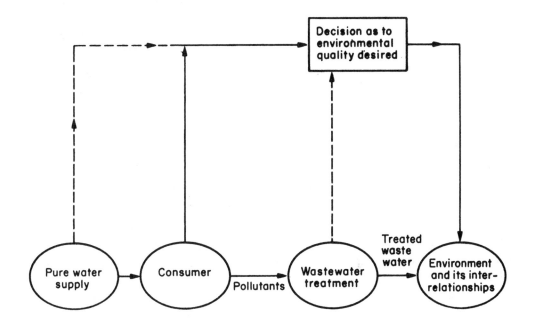

FIGURE 1. Model for Environmental Water Quality Control.

of approved treatment plants. All industries must comply with these discharge regulations, as well. Industries however, must pay for their own wastewater treatment by building on-site treatment plants or by paying user charges if they discharge to the local municipal treatment plant.

It is helpful if each discharger can utilize Figure 1 as a model in order to define the components, design a treatment plant for an acceptable cost which will provide the desired environmental water quality, and cope with instantaneous variations by implementing controls. Since the degree of wastewater treatment determines the extent of impact on the environment, both the citizen and the regulatory agency must agree that the greater the degree of wastewater treatment, the higher will be the cost and the smaller the impact of the discharge on the environment. The impact of treated wastewater discharges to the ecosystem is not strictly limited to the liquid phase of the biosphere, however. For example, since most drinking water purification or wastewater treatment processes convert pollutants to physically-removable solids, then the greater the degree of treatment, the greater will be the quantity of waste sludge solids generated. The resulting sludge will ultimately require disposal into the air or onto the land. Because the environmental impact of residual sludge disposal can not be eliminated, wastewater treatment alone can not be considered the ultimate solution for preventing undesirable impacts on the environment. Wastewater treatment will only reduce the direct impact of discharging pollutants to surface receiving waters.

B. Hydrological Considerations

The total mass of water associated with our planet is fixed and exists in various phases and locations, collectively referred to as the hydrological or water cycle. The world's supply of fresh water is extremely limited and results almost entirely from precipitation due to the evaporation of seawater. The major stages of the water cycle are as follows: precipitation, percolation, runoff, and evaporation. A schematic diagram of the water cycle is shown in Figure 2 [1].

It may not be obvious from this figure, that the major fraction of water which falls to earth as precipitation falls directly on water surfaces and is returned directly to the atmosphere by evaporation. Of the minor fraction which falls onto land masses, part is lost to the atmosphere by evaporation and transpiration from vegetation, part flows overland to receiving waters as surface runoff, and part enters the soil. The rainwater infiltrating the soil flows downward under the influence of gravity until it reaches the groundwater table to join the subterranean reservoir within the earth's crust. Most of the groundwater is eventually discharged at the ground surface through springs and outcrops, or it passes at or below the water level into streams and standing bodies of water. Ultimately, all the fresh water is discharged to the ocean where it is contaminated with salt and can no longer be used for most purposes.

In assessing the impact of potential discharges to surface waters, an initial consideration should involve an evaluation of the relative quantity of water in the various categories of the hydrologic cycle. Table 1 [2] indicates the total water budget of the world. As can be seen in this table, the total volume amounts to 1,360,000 cubic kilometers (326,000 cubic miles), but only a dramatically small fraction is freshwater. Only about 0.01% is in freshwater lakes and rivers, and only 0.61% in groundwater. The remaining water in the budget is contained in the oceans or in ice caps or glaciers—rather inaccessible.

The distribution of water in the conterminous U.S. is shown in

6

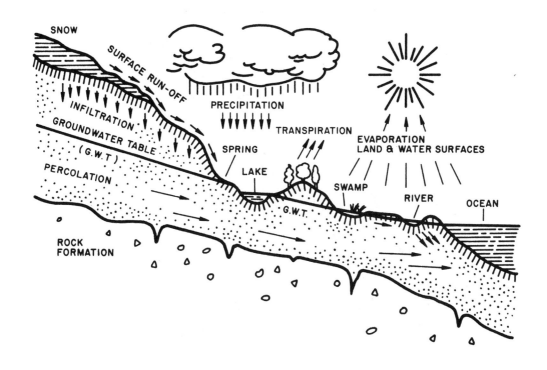

FIGURE 2. Schematic Diagram of the Water Cycle.

Table 2 [2]. Of particular significance are the relative volumes of water in freshwater lakes, streams and rivers, and groundwater compared to the total volumes of water in oceans, shown in Table 1. In addition, one should note that the ultimate impact of any discharge will depend to a large degree on the nature of the receiving water. Groundwaters are particularly vulnerable since they are utilized as a valuable drinking water source, and any pollution discharge to a groundwater aquifer will be retained in that supply for approximately 200 years [2].

The world's oceans possess the greatest volume and, therefore, the greatest capacity to dilute and absorb pollution discharges. Moving rivers and streams possess a moderate capacity to absorb pollution discharges and maintain an acceptable level of water quality. Freshwater lakes have demonstrated little capability to absorb pollution discharges without diminishing the water quality beyond acceptable levels for aesthetic or usage purposes.

Table 1

World Water Supply and Budget

Water item	Volume Cubic miles	Cubic kilometers	Available % Total Water
Land areas			
Freshwater lakes	30	125	0.009
Saline lakes and inland seas	25	104	0.008
Rivers	0.3	1.2	0.0001
Soil moisture and vadose water	16	67	0.005
Gr'ndwtr to 4,000 m (13000 ft)	2,000	8,350	0.61
Icecaps and glaciers	7,000	29,300	2.14
Sub-Total	9,100	37,800	2.80
Atmosphere	3.1	13	0.001
World oceans	317,000	1,320,000	97.3
Total	326,000	1,360,000	100
Annual evaporation			
From world oceans	85	350	0.026
From land areas	17	70	0.005
Total	102	420	0.031
Annual precipitation			
On world oceans	78	320	0.024
On land areas	24	100	0.007
Total	102	420	0.031
Annual runoff to oceans from			
Rivers and icecaps	9	38	0.003
Gr'ndwtr outflow to oceans	0.4	1.6	0.0001
Total	9.4	39.6	0.0031

TABLE 2

Distribution of Water in the Conterminous US

Water category	Area	Volume	Annual Circulation
	km²(mi²) × 10³	m³(mi³) × 10³	m³(ac − ft) × 10⁶
Frozen water			
Glaciers	0.51 (0.2)	0.52 (0.016)	0.0016 (1.3)
Ground ice	seasonal only and cannot be quantified		
Liquid water			
Freshwater lakes[a]	158.0 (61.0)	150.0 (4.5)	0.185 (150)
Salt lakes	6.7 (2.6)	0.45 (0.014)	0.0057 (04.6)
Ave streams		0.39 (0.012)	1.85 (1,500)
Groundwater			
Shallow	7,770 (3,000)	490 (15)	0.308 (250)
Deep	7,770 (3,000)	490 (15)	6.2 (5,000)
Soil moisture			
(3-ft root zone)	7,770 (3,000)	4.9 (0.15)	3.1 (2,500)
Gaseous water			
Atmosphere	7,770 (3,000)	1.45 (0.045)	6.2 (5,000)

a United States part of Great Lakes only.

C. Pollution of the Aqueous Environment

Because so many different substances can represent pollution—from pure elements to heterogeneous suspensions of complex substances—the measurement of pollution must be capable of representing any situation. One of the more common classifications of pollutants is by physical, chemical, and biological categories. Table 3 illustrates such a breakdown of significant pollutants in each category. Typically, if one must determine the pollution strength of a wastewater, analytical determinations should be performed for each pollutant shown.

Table 3

Common Measurements of Water Quality

A. Physical Measurements
1. Temperature
2. Color
3. Turbidity
4. Suspended solids

B. Chemical Measurements
1. Organic
   a. Biochemical oxygen demand
   b. Chemical oxygen demand
   c. Total organic carbon
2. Inorganic Fertilizing Elements
   a. Nitrogen
   b. Phosphate

C. Microbiological Measurements
1. Coliform bacteria
2. Fecal coliform bacteria
3. Standard plate count of total bacteria

In a broad sense, if one wishes to determine the impact of pollution discharge on the environment one can re-group all pollutants into one of two categories:

(1) organic pollutants which contain hydrocarbon fragments as a major portion of their composition, and

(2) inorganic pollutants which do not contain hydrocarbon fragments, e.g., dissolved minerals and salts which would exist in crystalline form in their pure states.

This broad distinction is convenient from an environmental perspective since there are two types of indigenous metabolic systems in nature; one system, used principally by animals, oxidizes organic materials which are in a reduced state; the other metabolic system, used principally by plants, has a remarkable capability to capture light energy in order to chemically reduce and sterically configure highly oxidized inorganic compounds. These two metabolic systems are called respiration and photosynthesis, respectively. In a biological sense, this distinction between plants and animals exists almost down to the unicellular level. At the unicellular level, these distinctions become blurred since some microorganisms can use both of these two opposing metabolic systems. But, simplistically, for modeling purposes, algae can represent the plant kingdom and bacteria can represent the animal kingdom. It is helpful to examine these two model organisms since, 1.) at this unicellular level of the ecosystem, any undesirable pollution discharge has a most direct and significant impact, and 2.) we can utilize these microorganisms in engineered biological treatment plants to utilize the biodegradable pollutants as food, and thereby degrade the wastewater. The goal of the engineer then becomes one of studying undesirable reactions occurring in nature and pre-empting nature by carrying out these reactions in an engineered reactor located at a biological treatment plant.

The metabolic systems of environmental respiration and photosynthesis are known as **heterotrophic metabolism** and **autotrophic metabolism**, respectively. Heterotrophic organisms require organic material as an energy source for growth, whereas autotrophic organisms utilize inorganic nutrients for growth. The heterotrophic class uses complex organic forms composed mainly of reduced forms of carbon, nitrogen, phosphorus, and sulfur, as well as trace elements. These foods or substrates can be broken down to produce energy, or they may be used to synthesize new cellular protoplasm. In general, these organisms respire, that is, they utilize oxygen to burn complex organic materials and produce carbon dioxide and water. The overall heterotrophic biochemical process known as **respiration** is represented by Equation 1.3 and illustrated in Figure 3.

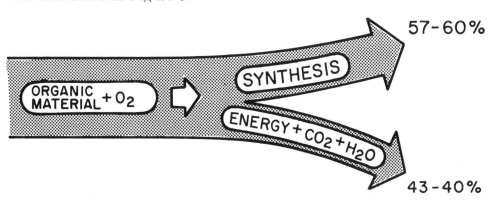

FIGURE 3. Schematic Representation of Heterotrophic Microbial Metabolism (Respiration).

$$\text{Organic Material} + \text{Oxygen} + \text{Microorganisms} \rightarrow$$
$$\text{New Microorganisms} + CO_2 + H_2O + \text{Energy} \tag{1.3}$$

In contrast, autotrophic organisms can convert inorganic materials into organic forms. These organisms utilize carbon dioxide (ultimately from the atmosphere) and highly oxidized inorganic forms of nitrogen, phosphorus, and sulfur as substrates. If the autotrophs metabolize these substrates via the process of **photosynthesis** in which solar energy is captured, they are called photo-autotrophs. The reaction is represented by Equation 1.4 and illustrated in Figure 4.

$$\text{Inorganic Nutrients} + \text{Sunlight} + \text{Microorganisms}$$
$$\rightarrow \text{New Microorganisms} + \text{Oxygen} + \text{Energy} \tag{1.4}$$

The photosynthetic process is important in many ways, for not only does it harness solar energy, but also it helps to remove the products of combustion and respiration. Also, it renews the oxygen supply to the respiring organisms and provides new organic matter for the whole food chain of respiring organisms, including man. All photo-autotrophs can use the process of photosynthesis during the day, thereby generating oxygen. At night, when no sunlight is available, they can respire, thus using the oxygen they have previously generated.

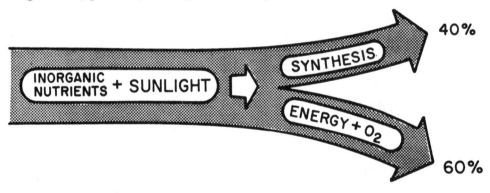

FIGURE 4. Schematic Representation of Autotrophic
Microbial Activity (Photosynthesis).

The photo-autotrophic and heterotrophic systems exist in a closed cycle or feedback loop. The end products from one type of metabolism provide input to the other type of metabolism; and the flow from one position within the loop to another is controlled by the material in the storage locations and the environmental conditions such as temperature, light, and mixing. A typical loop is shown is Figure 5.

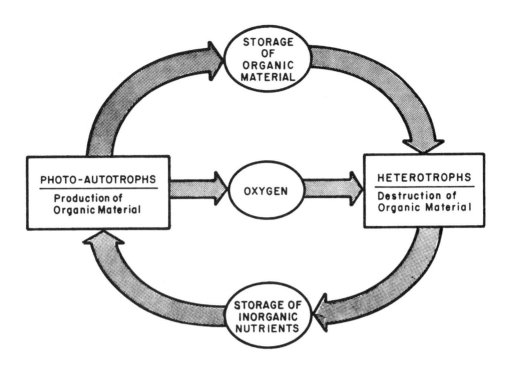

FIGURE 5. Heterotrophic/Photo-Autotrophic Balance
between Microorganisms.

The biosphere can be viewed as a macroscopic ecological system. In this system, a steady state between molecular production and destruction of organic material, as well as between production and consumption of oxygen, seems to be maintained. Hence a supply of oxygen is available in the atmosphere for respiring organisms, including man. In addition, the oceans serve as a sink for absorbing excess CO_2 produced from respiration and combustion. On a microscopic scale of a smaller ecological system, however, the balance between production and destruction may become easily disturbed. The introduction of either excessive organic or excessive inorganic nutrients can cause pollution impacts which perturb the balance between the heterotrophic and photo-autotrophic functions.

If excessive organic materials are introduced into the system, for example, heterotrophic activity will be in excess of photosynthetic oxygen release, and oxygen depletion will result. If, on the other hand, excessive inorganic nutrients are introduced into this system, autotrophic production rates will become larger than the rates of destruction of organic material by the heterotrophs and massive blooms of algae will occur. These two phenomena represent the two primary effects of pollution discharges to the environment and form the basis for evaluating the impact or perturbation of these discharges.

The final class of metabolic activity to consider is the case of heterotrophic activity when oxygen is absent. A specific class of heterotrophic organism exists which utilizes organic matter anaerobically and produces simple organic gases as end products. These organisms are called decomposers, and their anaerobic metabolic processes are represented by the sequential reactions in Equations 1.5 and 1.6 and are illustrated in Figure 6. The microorganisms carrying out Equation 1.5 are called acid formers, and those causing Equation 1.6 are called methane formers.

$$\text{Organic Material} + \text{Acid Formers} \rightarrow$$
$$\text{Volatile Acids} + \text{Energy} \tag{1.5}$$

$$\text{Volatile Acids} + \text{Energy} + \text{Methane Formers} \rightarrow$$
$$CH_4 + CO_2 + \text{Energy} \tag{1.6}$$

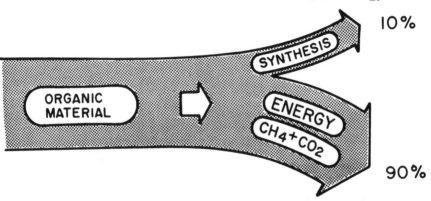

FIGURE 6. Schematic Representation of Anaerobic Heterotrophic Metabolism.

These microorganisms coexist with the autotrophs and aerobic heterotrophs. They are predominately scavengers which degrade the spent organic fauna and flora from the cycle illustrated in Figure 5. Figure 7 illustrates the interaction of the three types of microorganisms. Under normal conditions in a water body, the natural input to such a cycle shown in Figure 7 allows sufficient biological activity to provide food for higher forms of aquatic species in the food chain. The photoautotrophs, since they can produce organic compounds and living cells from inorganic nutrients, are called primary producers. Their products are utilized by heterotrophic organisms serving as primary consumers.

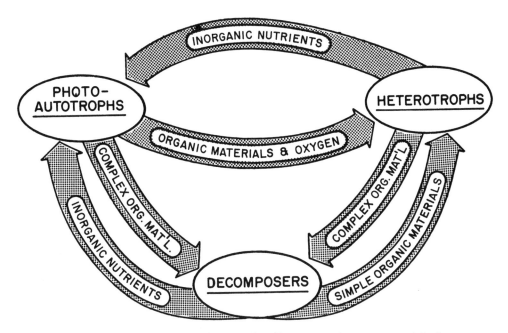

FIGURE 7. Inter-relationship between Heterotrophic/
Photo-Autotrophic/Decomposer Microorganisms.

The latter are then utilized by secondary consumers, etc., until a progression called the food chain can be traced. All of these living organisms die and are, in turn, degraded anaerobically by decomposer organisms.

If one considers the entire biosphere of land, water, and gaseous phases, this food chain must be described as a three-dimensional food web. Each ecological niche is occupied by many species which are competitive with each other and dependent on the next lower niche for survival. The food web is delicately balanced from the highest niche of humans down to the primary producers. The food web thus can be considered as a pyramid with each level dependent on the next lower level.

Any adverse effect upon the lower levels of the pyramid, where pollution discharges have their most direct impact, will have a rippling effect upwards to each higher level. An appreciable loss of energy occurs, however, at each trophic level of energy transformation, because each biological energy conversion is inherently inefficient. Figures 4, 5, and 6 illustrate the efficiency for each type of microorganism in converting food, or substrate, to new cells. Adverse effects from pollution discharges, felt most dramatically at the lower trophic levels, will be

15

successively dampened at each higher level. If these adverse effects reach the human niche, their magnitude would be serious.

The chemical formula of a microbial cell can be approximated as follows:

$$C_{106}H_{180}O_{45}N_{16}P_1$$

Other elements are also present, but only in trace amounts. The cells utilize various chemical compounds or substrates in this ratio as they synthesize new cellular protoplasm of this chemical formulation. If the required chemical compounds are not present in this ratio, one or more elements will be present in excess, and one element will be present at a minimum concentration. The latter element is considered to be the limiting substrate or limiting nutrient, and this reactant becomes a very important parameter in modeling the system, for it will determine the amount of growth. If either the organic or inorganic materials become depleted to a level which can not sustain the existing heterotrophic or autotrophic populations, then the population of one or both groups will drop to a level which the limiting substrate can sustain. Conversely, large pollution discharges into this cycle may stimulate an inordinately large population increase of one or both groups which will completely destroy previously-established ecological balances. Two important environmental effects are observed in the aquatic environment as a result of pollution discharges: (1) oxygen depletion due to organic pollutant discharges stimulating bacterial respiration, and (2) algal blooms due to fertilizing effects of discharged inorganic pollutants.

1.) Oxygen Depletion from Organic Discharge (DO Sag)

The introduction of organic pollution into a receiving water stimulates heterotrophic biological activity. Since heterotrophs utilize oxygen to respire via Equation 1.3, the dissolved oxygen concentration in the water is depleted. It is customary, therefore, to express the concentration of biodegradable organic pollutants as that concentration of OXYGEN required to biochemically degrade the organic material to carbon dioxide and water. This measurement of organic wastes is called BOD, the biochemical oxygen demand. The BOD is the most important measurement carried out on a wastewater for two reasons: (1) it indicates the oxygen depletion to be expected if untreated wastewater is discharged to the environment, and (2) it determines the required degree of wastewater treatment to be carried out before the wastewater is discharged so that minimal impact to the environment occurs. Historically, the impact of wastewater discharges to the environment has been determined by a consideration of dissolved oxygen dynamics in the receiving water. These dynamics combine an initial oxygen depletion and a re-aeration effect.

The pertinent dissolved oxygen relationships for developing the resultant oxygen sag curve are rather complicated. Gaseous, atmospheric oxygen continuously enters natural water systems due to transfer from the atmosphere according to Henry's Law, shown in Equation 1.7.

$$[\text{DO}]_{(\text{liq})} = K_H [pp]_{\text{O}_{2_{(\text{gas})}}} \qquad (1.7)$$

This equation indicates that the concentration of oxygen in the dissolved, liquid state is proportional by Henry's constant, K_H, to the partial pressure of oxygen in the atmosphere. The concentration of liquid, dissolved oxygen here is abbreviated as DO, enclosed in brackets, and specified by the subscript 'liq'. Equation 1.7 is a formal thermodynamic expression to indicate quantitatively the tendency for oxygen to be converted from the gaseous, atmospheric state to the dissolved, liquid state. The concentration of any parameter in the liquid state is expressed in units of mass/volume, usually in mg/ℓ or $gm/meter^3$. These units are appropriate for environmental engineering terminology because concentrations of most pollutants are very dilute.

The proportionality constant is called Henry's Law Constant. The value of this constant is an inverse function of temperature and is very small for oxygen, on the order of 46 mg/ℓ-atm at 20° C. Since the partial pressure for atmospheric oxygen is approximately 0.2 atm, the saturation, or maximum, concentration of DO at 20° C., as calculated from Equation 1.7, is 9.2 mg/ℓ. Unless the stream receives a perturbation in the form of another organic pollutant discharge, the stream can achieve and maintain this equilibrium saturation concentration. Appendix 1 provides the saturation concentration of DO at given temperatures and salinities. If the concentration of DO at any temperature and salinity rises above the given saturation concentration, we say that the system is 'supersaturated'. Conversely, if the concentration of DO drops below the given saturation concentration, we say that there is an oxygen 'deficit'. As previously described, the discharge of organic material to a receiving water results in a lowering of the dissolved oxygen concentration according to Equation 1.3. Figure 8 illustrates the case for a flowing stream.

The abscissa in Figure 8 can be expressed as time of flow or distance downstream. A hypothetical reduction in dissolved oxygen concentration from the saturation level due to an organic waste discharge at Point

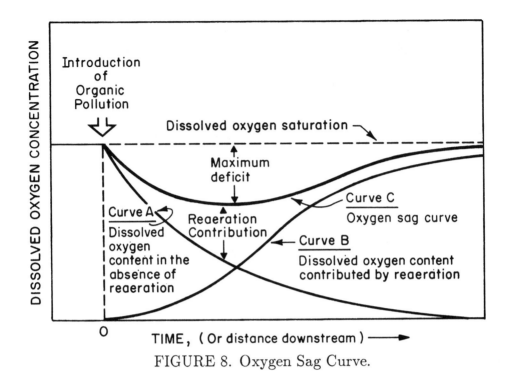

FIGURE 8. Oxygen Sag Curve.

0 time is shown as Curve A in Figure 8. This hypothetical reduction of dissolved oxygen is shown as if there were an absence of re-aeration capability so that Curve A never returns to the original saturation level. Since the flowing stream, however, is exposed to the atmosphere which contains approximately 20% oxygen, the resulting reduction in dissolved oxygen below the saturation concentration initiates a transfer of oxygen from the atmosphere to the stream as controlled by Henry's Law, and more dissolved oxygen is replenished to the water. This re-aeration effect is shown as Curve B in Figure 8. Curve A and Curve B are intimately related in natural systems, for the greater the level of oxygen depletion due to an organic discharge, the faster will be the rate of re-aeration.

Summation of Curves A and B results in Curve C, the oxygen sag curve. The sag curve expresses the oxygen deficit as the difference between the saturation concentration of DO and the measured concentration of DO.

$$\text{DO Deficit} = D = [\text{DO}]_{\text{saturation}} - [\text{DO}]_{\text{measured}} \qquad (1.8)$$

Curve C realistically describes the effect of oxygen depletion and reaeration in a moving stream due to the discharge of an organic waste-water into a moving stream. The mathematical formulation of this

18

concept was developed in 1925 by Streeter and Phelps [3]. Their formulation was based on the summation of the equations for Curves A and B, as follows:

1. OXYGEN DEPLETION: The rate of change in concentration of organic material, dy, expressed as units of BOD, with respect to time, dt, is equal to a rate constant, K_1, times the instantaneous concentration of organic material. Since the rate of change in concentration of organic material is decreasing with time, Equation 1.9 must contain a minus sign.

$$-\frac{dy}{dt} = K_1 y \tag{1.9}$$

It is important to note that $-dy/dt$, the rate of change in concentration of organic material is often called the rate of oxygen demand. And the rate of oxygen demand will equal the rate of change of oxygen deficit due to de-oxygenation shown in Curve A of Figure 8. Or,

$$-\frac{dy}{dt} = \frac{dD_{\text{Curve A}}}{dt} \tag{1.10}$$

2. RE-AERATION: The rate of change in oxygen deficit due to re-aeration, $dD_{\text{Curve B}}/dt$, equals another rate constant, K_2, times the dissolved oxygen deficit, D. Since re-aeration causes the rate of change in oxygen deficit to decrease, Equation 1.11 must also contain a minus sign.

$$-\frac{dD_{\text{Curve B}}}{dt} = K_2 D \tag{1.11}$$

The overall rate of change in oxygen deficit is a summation of the rate of increase in deficit due to oxygen depletion and the opposing rate of decrease in oxygen deficit due to re-aeration.

overall rate of change in deficit	=	rate of increase in deficit due to oxygen depletion	+	rate of decrease in deficit due to re-aeration

$$\frac{dD}{dt} = \frac{dD_{\text{Curve A}}}{dt} + \frac{dD_{\text{Curve B}}}{dt} \tag{1.12}$$

19

By substituting Equations 1.10 and 1.11 into Equation 1.12, the intermediate equation is obtained:

$$\frac{dD}{dt} = -\frac{dy}{dt} - K_2 D \tag{1.13}$$

And, finally, if the right hand side of Equation 1.9 is substituted into the first term on the right hand side of Equation 1.13, the final differential form of the Streeter-Phelps Equation is obtained:

$$\frac{dD}{dt} = K_1 y - K_2 D \tag{1.14}$$

Equation 1.14, a linear first order differential equation, can be rearranged into the classical form:

$$y' + Ay + B = 0 \tag{1.15}$$

where $y' = \frac{dD}{dt}$, $Ay = K_2 D$, and $B = -K_1 y$.

The detailed integration is presented in Appendix 2. The following expression results for the oxygen deficit, D, at any time or distance downstream from the discharge:

$$D = \frac{K_1 y_{mixed}}{K_2 - K_1} \left[e^{-K_1 t} - e^{-K_2 t} \right] + D_0 \left[e^{-K_2 t} \right] \tag{1.16}$$

where D_0 is the existing oxygen deficit at the point or time of discharge, and y_{mixed} is the concentration of biodegradable material immediately after mixing in the stream water.

The parameter y_{mixed} can be calculated from the concentrations of BOD in the wastewater discharge and that already in the receiving stream, using their respective volumetric flow rates. The subscripts w and s signify these sources of BOD, respectively. Again, y is the concentration of biodegradable material expressed in units of mg/ℓ, and Q is the volumetric flow rate in units of volume/time.

$$y_{mixed} = \frac{y_w Q_w + y_s Q_s}{Q_w + Q_s} \tag{1.17}$$

Equation 1.16 indicates that at any time t the impact from an organic discharge will depend on the existing deficit in the stream, D_0, the concentration of organic biodegradable material discharged and mixed with the river water, y_{mixed}, and the two rate constants, K_1 and K_2. These parameters can be measured or estimated in order to calculate D at any time t following discharge. Although Equation 1.16 is the formal integrand of Equation 1.14 in the base e system, engineers

20

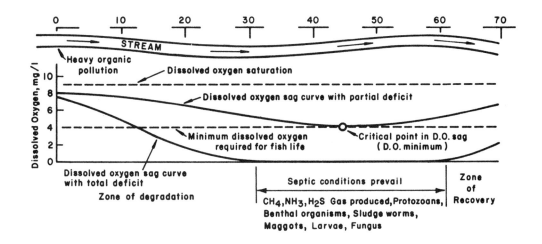

FIGURE 9. Two Oxygen Sag Cases.

often prefer to use the base 10 system. The notation used here will be big K for the base e system and little k for the base 10 system. The conversion of Equation 1.16 to the base 10 notation is as follows:

$$D = \frac{k y_{\text{mixed}}}{k' - k} \left[10^{-kt} - 10^{-k't}\right] + D_0\left[10^{-k't}\right] \qquad (1.18)$$

where, $k = K_1 /2.303$ and $k' = K_2 /2.303$.

Figure 9 illustrates two different cases of oxygen sag. One case illustrates a partial oxygen deficit and the other case shows a complete oxygen deficit. In the latter case, there is a stream stretch completely void of oxygen where anaerobic conditions will prevail, and there is a longer stream stretch in which the severe oxygen deficit prohibits the presence of fish. This figure also illustrates that regardless of the oxygen deficit value, if there is no additional discharge, in time there will be a recovery zone observed in which the dissolved oxygen concentration will begin to approach the saturation concentration.

It should be apparent from Figures 8 and 9 that a minimum occurs in the oxygen sag curve. This minimum occurs at a critical time, t_c, and is called the critical deficit, D_c. The critical time, t_c, can be found by differentiating Equation 1.16 or 1.18 with respect to t. The detailed differentiation is provided in Appendix 3 and results in the following equations for base e and base 10 functions, respectively:

21

$$t_c = \frac{1}{K_2 - K_1} \ln\left[\frac{K_2}{K_1}\left\{1 - \frac{D_0(K_2 - K_1)}{K_1 y_{mixed}}\right\}\right] \qquad (1.19)$$

$$t_c = \frac{1}{k' - k} \log\left[\frac{k'}{k}\left\{1 - \frac{D_0(k' - k)}{k y_{mixed}}\right\}\right] \qquad (1.20)$$

The resulting t_c value, calculated from Equations 1.19 or 1.20, can be substituted back into Equations 1.16 or 1.18 in order to calculate the corresponding D_c value. Example 1 following illustrates the use of these equations in base 10.

The Streeter-Phelps Equation, though only a simple model describing the effect of an organic discharge on a flowing stream, is very useful to provide a crude estimation of this impact.

EXAMPLE 1. A.) If a wastewater discharge with a residual concentration of BOD $= 8.0$ mg/ℓ and a volumetric flow rate of 0.44 m^3/sec is discharged into a stream already containing a BOD concentration of 1.0 mg/ℓ and a volumetric flow rate of 5.0 m^3/sec, what is the mixed concentration of organic matter just below the point of discharge prior to any oxygen demand being exerted?

ANSWER: Use Equation 1.17 to calculate y_{mixed} where Q is the volumetric flow rate and y is the concentration of BOD. The subscripts w for wastewater and s for stream are used:

$$y_{mixed} = \frac{y_w Q_w + y_s Q_s}{Q_w + Q_s}$$

$$y_{mixed} = \frac{\left[\left(8.0 \; \frac{mg}{\ell}\right)\left(0.44 \; \frac{m^3}{sec}\right) + \left(1.0 \; \frac{mg}{\ell}\right)\left(5.0 \; \frac{m^3}{sec}\right)\right]}{\left[0.44 \; \frac{m^3}{sec} + 5.0 \; \frac{m^3}{sec}\right]}$$

$$y_{mixed} = \frac{3.52 + 5.00}{5.44} \text{mg}/\ell$$

$$y_{mixed} = 1.6 \text{ mg}/\ell$$

B.)What will be the critical oxygen deficit, D_c and when will it occur if the existing oxygen deficit, D_0, is 1.2 mg/ℓ, $k = 0.35$ day^{-1}, and $k' = 0.2$ day^{-1} ?

ANSWER: First the critical time, t_c, must be found from Equations 1.19 or 1.20 as follows:

22

$$t_c = \frac{1}{k' - k} \log\left[\frac{k'}{k}[1 - \frac{D_0(k' - k)}{ky_{\text{mixed}}}]\right]$$

and, for this case,

$$t_c = \frac{1}{(0.2 - 0.35)} \log\left[\frac{0.2}{0.35}[1 - \frac{(1.2)(0.2 - 0.35)}{(0.35)(1.6)}]\right]$$

$$t_c = -\frac{1}{0.15} \log\left[0.57[1 - (-0.32)]\right]$$

$$t_c = (-6.67) \log(0.75)$$

$$t_c = (-6.67)(-0.12)$$

$$t_c = 0.8 \text{ days.}$$

The corresponding critical oxygen deficit, D_c, at t_c is then calculated from Equation 1.18, as follows:

$$D_c = \frac{ky_{\text{mixed}}}{k' - k}\left[10^{-kt_c} - 10^{-k't_c}\right] + D_0\left[10^{-k't_c}\right]$$

$$D_c = \frac{(0.35)(1.6)}{0.2 - 0.35}\left[10^{-(0.35)(0.8)} - 10^{-(0.2)(0.8)}\right] + (1.2)\left[10^{-(0.2)(0.8)}\right]$$

$$D_c = (-3.73)(-0.17) + (0.83)$$

$$D_c = 1.5 \text{ mg/}\ell$$

2.) Fertilization by Inorganic Nutrients (Eutrophication)

The discharge of inorganic nutrients, such as nitrogen and phosphorus, into a receiving water stimulates autotrophic biological activity. This system is shown in Figure 4. Most bothersome are the photoautotrophs, since these populations are quite visible, and large surges in population of these microorganisms are called algae blooms. The blooms begin in late spring, when the environmental conditions are conducive to growth, and continue through the summer. Growth extends as deeply into the water as the light necessary for photosynthesis can penetrate. The algae may reverse their metabolic mechanism at night when they can not photosynthesize and then respire via Equation 1.3. Hence the dissolved oxygen concentration in the upper layers of a water body can vary markedly throughout the 24-hour day. A

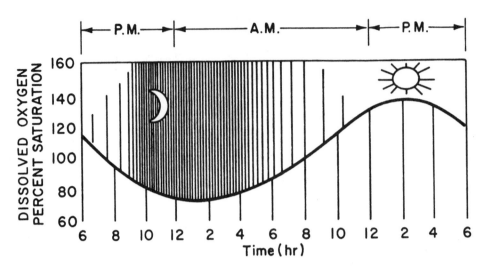

FIGURE 10. Diurnal Variations in DO Concentration.

typical result under these biologically productive conditions, called the diurnal variation in dissolved oxygen, is shown in Figure 10.

The introduction of inorganic nutrients into waters with long hydraulic detention times can be very dramatic. For such water bodies as lakes, reservoirs, impoundments, etc., can become stratified if they are deep enough. Stratification occurs when the depth precludes the energy of the wind from mixing the entire contents of the water body. In the summer months in a temperate climate, the upper waters will become warm and remain rather uniformly mixed due to the wind action. The cold water will be more dense and remain on the bottom. There is no mixing of the two zones, and they are separated by a boundary called the thermocline. A vertical temperature profile would indicate that the temperature decreases approximately $1°$ C. per meter of depth within the thermocline boundary layer. A typical summer temperature profile for such a body of water is shown on the right hand side of Figure 11.

No mixing occurs across the thermocline. The separate, distinct upper layer is called the epilimnion, and the comparable lower layer is called the hypolimnion. In stratified water bodies, there are only two time periods when the entire water contents are mixed; these are called the fall and spring overturns. The fall overturn occurs as the prevailing summer atmospheric temperatures begin to drop, resulting in a cooling effect on the warm top water layer. As the top layer cools, it becomes

24

more dense and sinks, causing the next warm layer to remain on top and undergo a sequential cooling and sinking process. This cycle is illustrated in Figure 12.

The cycle continues until the water has reached 4° C., when the water has achieved its greatest density. At this point, the entire volume of water is of uniform temperature and density. De-stratification occurs and the contents mix during the fall overturn. The resulting temperature profile is shown in Figure 11 on the center right. If the water cools below 4° C., the cooler water becomes less dense and rises to the top. If the temperature of the upper cooler water reaches 0° C., ice will form and the water body will again stratify—this time with a reverse temperature profile. The inverted winter temperature profile is shown in Figure 11 on the left. A similar overturn occurs in the spring when the ice melts and the warmer water sinks. When the entire water volume reaches 4° C., again, the contents mix in the spring overturn. This last temperature profile is shown in Figure 11 at left center.

During summer stratification, vast changes occur in the lake as a result of pollution. The algal blooms resulting from inorganic pollution discharges may cause the upper layers to become alternately supersaturated with oxygen during daily periods of photosynthetic activity, via Equation 1.4, and totally void of oxygen during the night when the algae respire, via Equation 1.3. The only oxygen replenishment at night occurs through diffusion from the atmosphere at the surface as dictated by Henry's Law. This dissolved oxygen is utilized by the respiring microorganisms almost as fast as it can diffuse into the water. But because no mixing occurs across the thermocline, no oxygen replenishment can occur in the hypolimnion. Respiring organisms in this layer will completely exhaust the dissolved oxygen and anaerobic conditions will exist at the bottom. A daytime vertical dissolved oxygen profile of such a polluted lake is shown in Figure 13.

Throughout summer periods of heightened biological activity, approximately 4% of the photo-autotrophic microorganisms (algae) and the heterotrophic microorganisms (bacteria) produced in the epilimnion die and settle through the thermocline daily and form a layer on the bottom muds. Here they serve as substrate for the bottom anaerobic decomposers, according to Equations 1.5 and 1.6. Again, the degree of pollution present and the resulting oxygen depletion will dictate whether the hypolimnion will be aerobic or anaerobic. Usually under polluted conditions, the bottom water becomes void of oxygen and the anaerobic decomposition reaction of Figure 6 prevails. A cycle between the autotrophs and aerobic heterotrophs in the epilimnetic layer

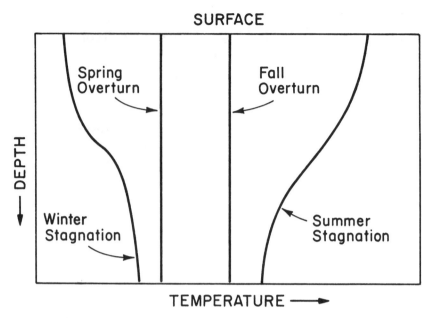

FIGURE 11. Temperature Profile in a Stratified Lake.

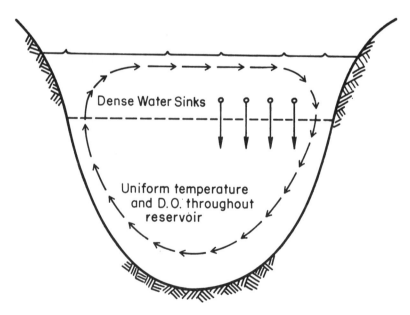

FIGURE 12. Diagram of Lake Turnover.

and the anaerobic decomposing microorganisms in the hypolimnion is thus established in the stratified lake as shown in Figure 7. Figure 14 summarizes these relations and their effect on water quality.

During the winter stratification period, anaerobic decomposition of the previous summer's crop of dead microorganisms occurs. Since the rate of decomposition is decreased at lower temperatures, rarely can a year's accumulation of biologically-produced organic matter be degraded by the anaerobic decomposers in the bottom muds if pollution inputs are excessively large. Instead, the majority of residual organic debris accumulates in the sediments until it is re-suspended and re-circulated throughout the entire lake volume during the next spring overturn. These materials are then present to initiate the annual cycle of bacteria and symbiotic algae blooms shown in Figure 5. The cycling of pollutants between the water and the biological species in a closed system is shown in more detail in Figure 15.

The cycling of pollutants is referred to as a biogeochemical cycle and is particularly significant in the case of nutritional elements such as phosphorus. As can be seen in Figure 15, such an element is conserved in the cycle and does not leave the system, except for the small concentration which is slowly washed out due to hydraulic inputs and out-flows.

This cycle continues year after year, and its magnitude will be increased proportionately with increasing pollution discharges to the lake. Unfortunately, the cycle will never be reduced in magnitude unless the lake is treated to remove the pollutant materials or until the discharges are halted and previous concentrations are reduced by natural hydraulic flushing.

In order to halt the biogeochemical cycle, such treatment schemes as hydraulic flushing, removal or covering the bottom muds, and harvesting of aquatic plants have been proposed. If the polluted water body is left untreated or continuously receiving pollutant discharges, however, the water quality deteriorates, the biological activity increases, the bottom accumulations become deeper and deeper, and ultimately the lake becomes a swamp. At that point the lake ceases to be a water resource.

This process of aging is a natural one and is known as eutrophication. This term stems from the Greek word *eutrophos*, meaning 'well-nourished' and refers to the increased productivity which develops in the lakes as they become richer in nutrients. The following characteristics can be used to classify a lake as to its degree of eutrophication:

The lake extinction proceeds through three limnological classifications:

27

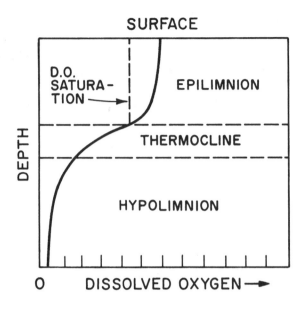

FIGURE 13. DO Profile in a Stratified Lake.

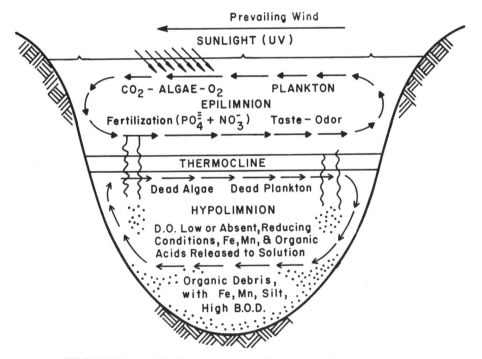

FIGURE 14. Biological and Chemical Effects in a
Eutrophic Lake.

28

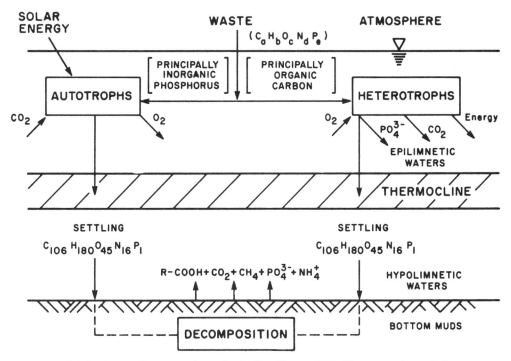

FIGURE 15. Biogeochemical Cycling of Pollutants in a lake.

    (1) water quality,
    (2) biological activity, and
    (3) depth of bottom muds

1. *oligotrophic*, or young, with low nutrient content, little biological activity, and a clean bottom sediment,

2. *mesotrophic*, or middle-aged, with modest amounts of the three parameters, and,

3. *eutrophic*, or aged, with large concentrations of nutrients, excessive biological activity, and large volumes of organic sediments on the bottom.

Figure 16 illustrates these stages and the extent of their biological cycles [5].

Although eutrophication is a natural environmental process, normally tens of thousands of years are required to complete the process. Human activities, however, have so accelerated the process that lakes sometimes become eutrophic in a matter of years. A graph projecting

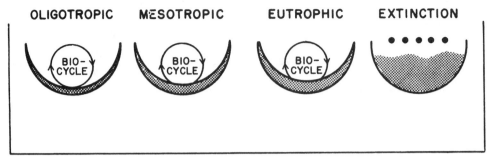

OLIGOTROPIC   MESOTROPIC   EUTROPHIC   EXTINCTION

TIME ⟶

FIGURE 16. Transition Stages of a Lake.

the time relationship for natural eutrophication and acceleration due to man's activities is shown in Figure 17. In this figure, the parameter used to illustrate the relative rates of eutrophication is measured in terms of the biological productivity on the ordinate. Actually, the nutrient content, the biological productivity, or volume of bottom accumulation could be plotted as an index of the degree of eutrophication. If the number of uses for the water is considered with respect to the degree of eutrophication, then, when the lake is very young, the water quality is very high and it might be used directly for human consumption. And conversely, as the lake ages, the water quality will deteriorate so that first, it will no longer be suitable for drinking, then no longer suitable for contact sports, then no longer suitable as a fishing resource, until ultimately, all its uses will be lost.

REFERENCES

[1] G. M. Fair, J. C. Geyer, and D. A. Okun, *Water and Wastewater Engineering,* **Vol. 1,** Wiley, New York, 1968.

[2] D. K. Todd, *The Water Encyclopedia,* Water Information Center, Port Washington, NY, 1970.

[3] H. W. Streeter and E. B. Phelps, *Public Health Bulletin,* **No. 14** (1925).

[4] C. N. Sawyer, *Journ. Water Pollut. Control Fed.,* **38,** 737 (1966).

HOMEWORK PROBLEMS

1. Identify the three physical phases of the biosphere and give an example of a natural component in each phase. List as

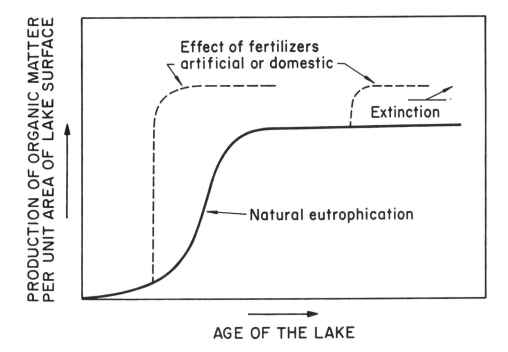

FIGURE 17. Relationship of Lake Eutrophication
and Productivity (Natural and Human Impact).

many pollutants (or classes thereof) which you can think of that
are routinely discharged to each phase of the biosphere.

   2. Define the following terms:
      a. steady state
      b. absorptive capacity
      c. equilibrium state

   3. Define the following terms:
      a. heterotrophic organisms
      b. photo-autotrophic organisms
      c. symbiotic relationship

   4. Briefly describe the hydrologic cycle.

   5. Explain how respiration and photosynthesis interact.

   6. Why is the concentration of biodegradable organic material expressed in units of dissolved oxygen?

7. Define the following terms and indicate the symbols used for these parameters:
  a. dissolved oxygen
  b. dissolved oxygen saturation
  c. oxygen deficit

8. The existing oxygen deficit of 1.0 mg/ℓ is observed in a river with a linear flow velocity of 0.18 m/sec. After discharge of a waste effluent into this river, $y_{mixed} = 5.0$ mg/ℓ, $k = 0.15$ day^{-1}, and $k' = 0.33$ day^{-1}, calculate the oxygen deficit, D, after 2 days.

9. If the dissolved oxygen concentration in Problem 8 initially is 8.5 mg/ℓ, what is the DO concentration 2 days after discharge?

10. What is the critical time at which the largest deficit occurs in Problem 8?

11. What is the resulting deficit at the critical time?

12. If state regulations required a residual DO concentration of 4.0 mg/ℓ at the critical time after discharge, would a discharge permit be issued for the effluent in Problem 8?

13. Knowing the linear flow velocity from Problem 8, how far downstream will the critical deficit occur?

14. Define the following terms:
  a. diurnal effect
  b. stratification
  c. eutrophication
  d. oxygen profile
  e. temperature profile

COMPUTER PROBLEM

Calculate the largest concentration of y_{mixed} which would be permitted for the conditions described in Problems 8 – 12. HINT: Determine the maximum D_c permitted. Then carry out a trial and error solution to find the maximum y_{mixed} with unknown t_c. Use Equations 1.18 and Equation 1.20, for example.

Chapter 2

POLLUTANT CHARACTERISTICS

A. Wastewater Components and Impacts

One or more of the following immediate reactions may occur as a result of discharging pollutants into the aquatic environment:

1. The components may react chemically with other dissolved or suspended ions or compounds.

2. The components may be taken up biologically and therefore stimulate the metabolic activity of any indigenous organisms.

3. The components may remain in solution, thus increasing the concentration of total dissolved solids in the system.

The principal elements that would be released to the environment as a result of the discharge of pollutants are: carbon, nitrogen, phosphorus, and sulfur. Each of these elements can exist in numerous forms, and in each case the potential transformations which can occur exist within a natural cycle for that element. The cycles for these elements are based on the chemically- and biologically-induced transformations which are favored for each of these elements and, in many cases, will include portions of the general biological cycles previously illustrated in Figures 5 and 7.

Our environment is composed of numerous biological and chemical cycles of this nature and the quality of our environment is, in fact, determined by these cycles. None of these cycles is singularly independent, but rather is mutually inter-dependent, the outputs of one cycle serving as inputs to other cycles. In a closed ecological system such as the biosphere or a segment of the biosphere, there must be a balance between the inputs and outputs of each of the cycles as there is neither an infinite sink for, nor source of, any component. The quality of the environment thus is established by controlling the magnitude of the respective cycles and maintaining the proper balance between each of the cycles. Figure 18 illustrates the inter-relationship between the various cycles as a system wherein each cycle can inter-relate with a number of other cycles.

Although such a model is greatly simplified, it demonstrates the interdependence of the factors which serve to establish a given environmental quality. Furthermore, such a model clearly illustrates how a given pollutant can affect numerous aspects of the environment. In order to determine accurately the impact of a given pollutant on the environment, it is desirable to understand as many of the fundamental cycles as possible.

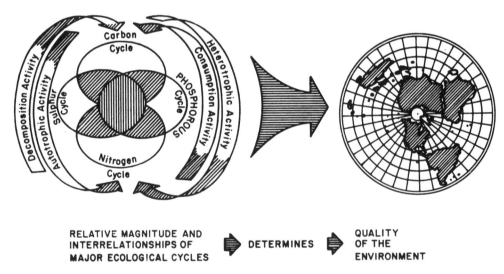

RELATIVE MAGNITUDE AND INTERRELATIONSHIPS OF MAJOR ECOLOGICAL CYCLES ➡ DETERMINES ➡ QUALITY OF THE ENVIRONMENT

FIGURE 18. Schematic Diagram of Elemental Cycles and their Effects on Environmental Quality.

The magnitude of such an analysis may be appreciable and, unfortunately, there are currently many gaps in the available data. Thus, as previously discussed, the impact of pollution frequently is still measured solely in terms of simplifying yardsticks such as heterotrophic activity (i.e., oxygen depletion) and photo-autotrophic activity (i.e., primary productivity).

One proposed method attempting to quantify numerous environmental parameters as they relate to water quality is the development of a Water Quality Index [1]. Such an index might range from 0 (poor) to 100 (excellent) and be based on a proportionate weighted evaluation of various water quality parameters such as dissolved oxygen, biochemical oxygen demand, fecal coliform, pH, nitrate, phosphate, temperature, turbidity, and dissolved solids. Development of such a Water Quality Index would allow for immediate recognition of a water quality, similar, for example, to the immediate recognition of a number on the Richter scale with earthquake intensity.

The most precise assessment of pollution impact on the environment can be made, however, only by determining the effect of a given discharge on a specific site. Although certain threshold levels of various pollutants can be tolerated, the excessive discharge of a given pollutant should be eliminated to reduce resulting imbalances of natural cycles.

A summary of major cycles follows, and the environmental impact of each element is discussed.

1. Carbon

The potential carbon input from environmental discharges can be significant. The element carbon occupies an intermediate position between electropositive and electronegative elements in the Periodic Table of Elements. Carbon can, therefore, exist in several pure elemental forms as well as in highly reduced or oxidized forms. It can exist in solid, liquid, or gaseous phases, in pure or combined states and it can circulate through the organic, inorganic, living, and nonliving cycles. Carbon is the most ubiquitous element within living protoplasm, built from the carbon skeletal matrix of carbohydrates, proteins, and lipids, and, the carbon skeleton of intermediate metabolites provides the primary fuel for combustion processes in respiring organisms.

The earth's crust contains many forms of inorganic carbon as well as the great volume of degraded forms of previously living carbon. The 2% of the earth's atmosphere, which is composed of inorganic carbon dioxide, serves as a source of carbon substrate for photo-autotrophic organisms and a sink for this metabolite from respiring organisms. It is this delicate concentration balance of atmospheric carbon dioxide which is being threatened increasingly by excessive energy consumption in the developed countries. At the present time, the increased carbon dioxide production from more numerous combustion systems or from increased activity of respiring organisms can be absorbed by the world's oceans. If the seas' absorptive capacity is exceeded, however, it is postulated that the concentration of carbon dioxide in the atmosphere will increase, which will result in an insulating effect or 'greenhouse effect' which might serve to raise the prevailing ambient temperature. On the other hand, if the world's carbon dioxide production can be sufficiently curtailed so that the numerous carbon forms can continue to remain in balance with one another, then the various inputs and outputs can flow continuously from one form to another. The interchanges characteristic of the carbon cycle are illustrated in Figure 19 [2].

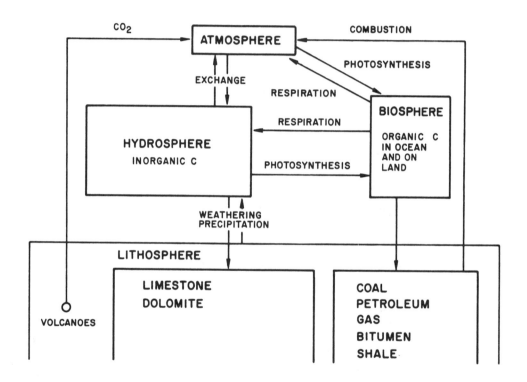

FIGURE 19. Circulation of Carbon in Nature.

The principal inputs from wastewater discharges to this cycle originate from the organic fraction of body wastes. This wastewater pollutant, if left untreated, enters into the organic carbon position and from there can be transported throughout the cycle. The other minor carbon-containing components such as biodegradable organic chemicals could also enter into the cycle at the same position.

The inorganic carbon component in wastewater would originate from chemical discharges. Inorganic carbonate is removed only slightly by photo-autotrophic microorganisms in a biological waste treatment plant and passes through virtually intact. The untreated carbonate, released in the treatment plant effluent, enters the carbon cycle at the inorganic carbon position, stimulates the growth of photo-autotrophic microorganisms, and is subsequently transported throughout the cycle.

2. Nitrogen

The element nitrogen also exists in many forms in all phases of the biosphere. It exists in elemental form as a diatomic gas in the atmosphere. It also occurs in various inorganic forms with oxidation states from the most reduced compound ammonia, NH_3, to the most oxidized form nitrate, NO_3^-.

All of these oxidation states fill important roles in the interchange between non-living and living forms. Domestic wastewaters have large fractions of nitrogen-containing urea. Nitrogen-using organisms can break this and similar organic nitrogen compounds down to inorganic ammonia. The discharge of ammonia subsequently imposes an oxygen demand on a receiving water when it is converted biochemically to nitrogen oxides or cellular components. These transformations from inorganic to organic forms and reversion back to inorganic forms occur primarily by microbial action. Energy is required to convert each form stepwise throughout the biological pathways.

Energy in the form of lightning may convert atmospheric nitrogen gas into ionic forms, which are then washed from the air by rainwater and introduced into the ecosystem. The inter-relations and conversions of different nitrogen forms are illustrated in Figure 20 [3]. One might note how chemical fertilizers containing ammonium salts would enter and affect this cycle. It is now estimated by EPA that ammonium inputs from storm water runoff of agricultural lands create larger oxygen demands on natural receiving waters than any other source.

3. Phosphorus

Phosphorus exists almost always in its most highly oxidized state, i.e., phosphate ion. This fact does not lead to an oversimplified role for phosphorus in the environment; instead, this fact complicates the case considerably since the phosphate ion undergoes only minor chemical and phase transformations in order to become transported throughout the cycle. Appreciable amounts of organic phosphates are excreted in human wastes and contribute significantly to municipal wastewaters;

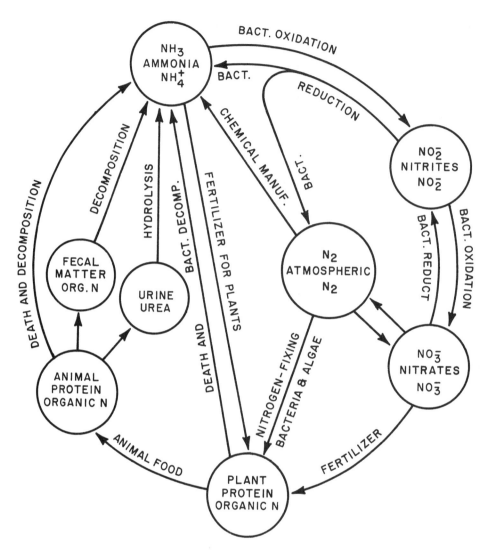

FIGURE 20. Circulation of Nitrogen in Nature.

these organic phosphates are easily hydrolyzed to inorganic phosphate. Poly-phosphates, which are inorganic condensed phosphates, can be a significant component of synthetic detergents.

Inorganic phosphate is assimilated by living cells, and converted into high-energy phosphate bonds which later serve as energy sources for thermodynamically-unfavorable biochemical reactions required for metabolism. For respiring organisms, the respiration reaction in Equation 1.3 can be separated into two steps, as illustrated in Figure 21.

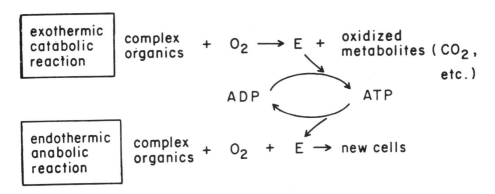

FIGURE 21. Energy Exchange in Respiration Reaction.

Here, the catabolic respiration reaction produces energy which is captured by adenosine diphosphate in a 'high energy bond' to yield Adenosine Tri-Phosphate, as shown in Figure 22.

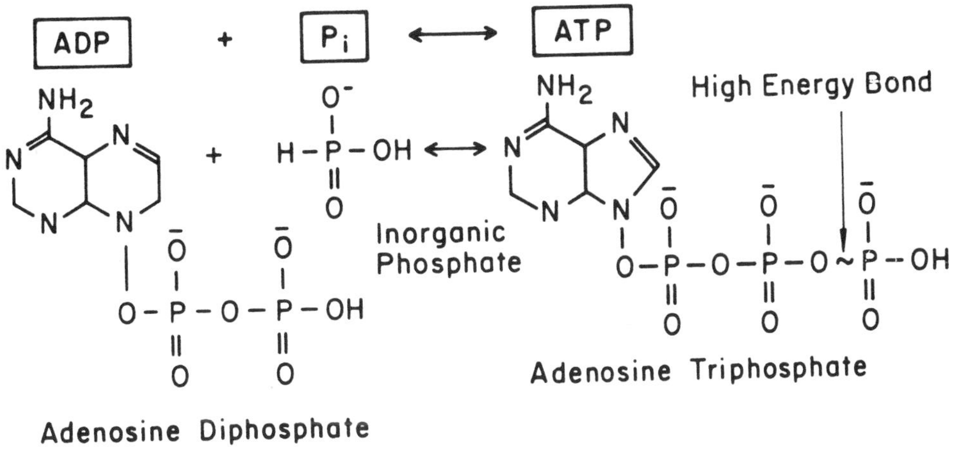

FIGURE 22. ADP/ATP Energy Interchanges.

The final phosphate added to the Adenosine Di-Phosphate in a 'high energy bond' subsequently can be broken to produce energy required by the anabolic respiration reaction shown in Figure 21. The Adenosine Di-Phosphate produced by the anabolic reaction is then ready to undergo the same cycle over again. The ADP-ATP reaction is 'coupled' to the anabolic-catabolic reactions as illustrated Figure 21. The reversible ADP/ATP reaction is shown in both shorthand form and in detail in Figure 22.

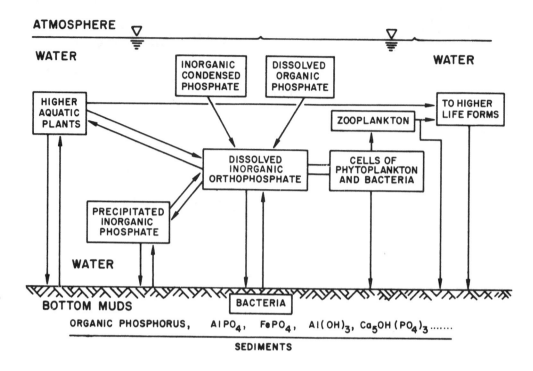

FIGURE 23. Phosphorus transformations.

In addition, inorganic phosphate ions may become complexed with metal ions such as calcium, iron, or manganese and remain in solution, or phosphate ions may be adsorbed on the solid surfaces of clays or muds present in the bottom of surface water bodies. Inorganic phosphate under certain pH conditions may also precipitate with cations such as calcium, iron, manganese, or aluminum to form solid sources of phosphate. The phosphate cycle illustrated in Figure 23 shows these transformations [2].

Phosphate components formerly comprised an appreciable fraction of commercial detergents in the past and caused a significant input to the environment because conventional biological wastewater treatment does not remove these ions when present in excess of cellular requirements. Evaluation of the exact impact of this particular input on the environment has been the subject of extensive controversy over the years, particularly with regard to phosphate's potential nutrient-limiting role in the environment. Analysis of the elements necessary for algal growth, as shown in Figure 4, indicates the following: (1) carbon dioxide is available from either waste products of aquatic respiring organisms or the atmosphere, (2) nitrogen is available either from storm water runoff of agricultural and urban lands or from waste products of respiring organisms; in addition, certain algal species are capable of 'fixing' (or utilizing) atmospheric nitrogen, and (3) hydrogen and oxygen are available as ions or in the dissolved molecular form in aqueous systems. In this context, phosphate and requisite trace-metal concentrations are generally considered as the more probable limiting elements with respect to algal growth.

Thus, the argument has frequently been given that the phosphate content of synthetic detergents has served to increase the frequency of algal blooms in the nation's surface waters. A critical analysis of each particular case poses the following questions:

1. Is phosphate the limiting nutrient, which, if absent, would preclude excessive algal blooms?

2. How much of the phosphate present in the environment resulted from detergent input?

3. Are the proposed phosphate substitutes more harmful to the environment than phosphates?

Various studies have yielded different answers to each of these questions. This is simply because each water body exhibits its own characteristics and no two are alike. Thus no *a priori* statement can be made that phosphate discharges are either harmful or not harmful to the environment. This decision can be made only after performing a technical evaluation of all the inputs to and outputs from a particular water body and determining its inherent biological and chemical characteristics.

4. Sulfur

The element sulfur can exist in pure elemental form or in intermediate oxidation states from the most reduced form, hydrogen sulfide, H_2S, which can cause significant odor problems, to its most highly oxidized state, sulfate, SO_4^{-2}. The latter form causes a cathartic effect in the human body and, therefore, its recommended concentration in waters meant for human consumption is limited to 250 mg/ℓ [4]. The former reduced oxidation state is utilized by living systems to form organic sulfides which are important function-determining and structure-determining groups in proteins. The inter-related transformations are illustrated in Figure 24 [3].

The major input to this cycle from wastewater discharges originates from human wastes and inorganic chemicals commonly used in both households and industries. Only a small fraction of the discharged concentration is removed by biological wastewater treatment plants. The major fraction is released to the environment and the significance of this input has not been investigated thoroughly. Sulfate ion has appeared to have a variable effect on different fish species and lower fauna species whereas sulfide ion is highly toxic to fish.

5. Miscellaneous Elements

Many commonly-used products in households and industries contain small concentrations of miscellaneous elements which enter wastewaters. Since no appreciable amount of these components is removed by biological wastewater treatment processes, they tend to accumulate in the environment and increase the existing level of total dissolved solids in waters. The existing water supplies do not have an infinite capacity to dissolve solids and, if advanced wastewater treatment controls are not implemented to remove these components, the drinking water tolerances for various elements will be frequently exceeded.

The U.S. Public Health Service has set a limit for total dissolved solids in drinking water supplies not to exceed 500 mg/ℓ [4]. Currently, this limit is exceeded only in certain areas of the United States. In the future, though, as multiple use and reuse of fresh waters becomes more prevalent, concentrations of these components can increase to excessive levels.

Consequently, the public must decide according to models such as the one shown in Figure 1 whether to require advanced treatment for removal of these inorganic ions and thus achieve a purer environment and, ultimately, purer drinking water, at increased cost. Conversely, the inorganic ions can be released to the environment with a corresponding sacrifice in environmental quality and drinking water quality.

The inorganic ions in question consist of carbonates, chlorides, silicates, borates, perborates, sodium, and heavy metals. The carbonates and sulfates which have been discussed previously and the remaining inorganics deserve individual mention. Chlorides impart a brackish taste to water and, since they normally are untreated in wastewater treatment plants, their presence is used as an indicator of pollution inputs in freshwater systems. Their concentration is limited to 250 mg/ℓ in domestic drinking water supplies [4]. Chlorides are not involved in any biological cycle as such, but their negative ionic charge provides electroneutrality for living organisms containing high concentrations of

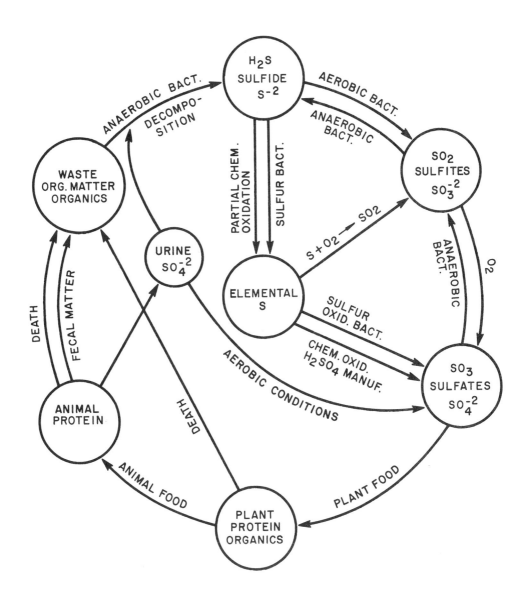

FIGURE 24. Circulation of Sulfur in Nature.

positively-charged ions in their cells.

Silicates serve as a macronutrient for diatomaceous algae. This type of algae is very prevalent in the sea and serves to convert soluble silicate forms into solid skeletal structure in their cells. The cells may be free-floating throughout their lifespan and then settle to the bottom sediments of the water body. The dead silicaceous algae may then become part of the permanent bottom muds or may be degraded by decomposer bacteria which resolubilize the silicate. Thus the biological silica cycle transports any silicate through the ecosystem.

The borates and perborates are utilized in trace amounts by most green algae and plants. Any significant boron input to natural waters would remain in excess of the concentration removed by biological treatment, since it is needed in such trace amounts. The excess borates and perborates are toxic to certain plants in concentrations greater than 2 mg/ℓ and are thought to be innocuous to plants and animals at concentrations less than 0.1 mg/ℓ.

Sodium ion concentrations can become significant, from a public health aspect, due to their potential deleterious effects on people with circulatory ailments. For this reason, the Environmental Protection Agency in reviewing the 1962 U. S. Public Health Service Drinking Water Standards has recommended that sodium concentration in drinking water should be reported to the physicians in the utility service area for the benefit of patients requiring low-sodium diets. Sodium ion is also thought to enhance the ability of algae to assimilate luxury amounts of phosphate in natural systems [5]. This enhancement, in specific cases, may lead to increasing rates of eutrophication.

B. Pollutant Measurements and Characteristics

1. Drinking Water Characteristics

Humans interrupt the natural water cycle illustrated in Figure 2 when water is removed from the environment or discharged to the environment. Each successive use of the water will degrade its quality because the expense to purify the water 100% is prohibitive. Everyone, however, is looking for a 'pure' water source which requires little treatment for safe human consumption or reliable industrial use.

Drinking water is supplied by a municipality to citizens within the municipal boundaries. Households outside incorporated limits must provide their own water supply. Sometimes there is a private water supply company. The 1974 Clean Drinking Water Act stipulates that both public and private water supply companies serving more than 15 households must provide satisfactory drinking water which complies with U. S. Public Health Drinking Water Standards shown in Table 4.

These U. S. Public Health Drinking Water Standards have not been revised and approved since 1962. Interim modifications have been adopted, however, and are indicated in Table 4, even though they are not 'approved'. The standards are divided into three major categories,

44

Table 4
Modified 1962 US Public Health Service Drinking Water Standards

1. BACTERIOLOGICAL:
Only one bacterium per 100 ml can be present.

2. PHYSICAL:

Characteristic	Not Greater Than:
Turbidity	1 Turbidity Unit*
Color	15 Color Units*
Odor	3 Odor Units*

*These arbitrary and subjective units are defined in approved tests.

3. CHEMICAL:

	Limits set for reasons of	
	Aesthetics	Health
		mg/ℓ
**Arsenic (As)		0.05
**Barium (Ba)		1.0
**Cadmium (Cd)		0.01
**Chromium (Cr)		0.05
Cyanide (CN)		0.2
Fluoride (F)		1.4 - 2.4
**Lead (Pb)		0.05
**Mercury (Hg)		0.002
Nitrate Nitrogen (NO_3^-, as N)		10.0
**Selenium (Se)		0.01
**Silver (Ag)		0.05
***Endrin		0.0002
***Lindane		0.004
***Methoxychlor and 2,4-D		0.1
***Toxaphene		0.005
***2,4,5-TP		0.01
***Ra^{226} and Ra^{228}		5 pCi/ℓ
***α particles		15 pCi/ℓ
****Total trihalomethanes (TTHM)		0.1
Chloride (Cl)	250.0	
Copper (Cu)	1.0	
Iron (Fe)	0.3	
Manganese (Mn)	0.05	
Methylene blue active substances	0.5	
Sulfate (SO_4^{-2})	250.0	
Total dissolved solids	500	
Zinc (Zn)	5.0	

** Toxic Heavy Metals
*** Adopted as interim standard in 1976, but not approved by EPA
**** Adopted as interim standard in 1980, but not approved by EPA

and when a municipally treated water supply can meet these standards in all categories, it can be distributed throughout the municipality in a pressurized water distribution system. After the water is used, its quality has been degraded and it must be collected in a separate piping system and sent to a wastewater treatment plant where it is treated prior to its discharge back to the environment.

2. Wastewater Characteristics

The strength of wastewater is measured principally by the concentration of its biodegradable carbon and nitrogen, which cause an oxygen demand in the receiving water. Another important measure of pollution is the turbidity, measured as suspended solids.

A. Oxygen Demand

Biodegradable carbon and nitrogen are measured by the BOD analysis.

BOD = Biochemical Oxygen Demand, or the quantity of dissolved oxygen required to metabolize the biodegradable material in a liquid sample.

In a batch test, the continual decrease of DO can be measured as shown in Figure 25.

As can be seen in this figure, the depletion rate of DO is proportional to the degradation rate of biodegradable material. Therefore, the concentration of DO can be measured and SET PROPORTIONAL to the concentration of biodegradable material. Since each parameter declines at a proportional rate in a batch test, the measured DO concentration is used as an indicator of the concentration of remaining biodegradable material.

The rate of decrease in DO is a first order decay rate, comparable to radioactive decay studied in chemistry class. The rate of decrease in biodegradable material is first order since it is proportional to the rate of DO decrease. Or, if the concentration of remaining biodegradable material is L, then

$$-\frac{dL}{dt} = K_1 L \qquad (2.1)$$

Remember that a decreasing concentration of biodegradable material requires that the left hand side of Equation 2.1 is negative. Grouping terms:

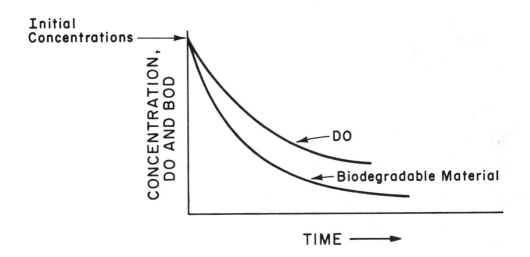

FIGURE 25. Decrease in Residual DO and Residual
Biodegradable Material with Time.

$$\frac{dL}{L} = - K_1 \, dt \qquad (2.2)$$

Integration yields the following expression,

$$\ln \frac{L}{L_0} = - K_1 t \quad \text{or}$$
$$\log \frac{L}{L_0} = - kt \qquad (2.3)$$

In exponential form, the integrand may be expressed as follows:

$$L = L_0 \left[e^{-K_1 t} \right] \quad \text{or}$$
$$L = L_0 \left[10^{-kt} \right] \qquad (2.4)$$

where $k = K_1 / 2.303$, and L_0 is defined as the initial concentration of biodegradable material. Although Figure 25 indicates DO and the corresponding concentrations of residual biodegradable material are continuously measurable, this is not true under the standard conditions

47

used to carry out the BOD test. A description of this procedure will follow later and you will see that the DO is measured only after one 5-day incubation period, and the initial concentration of BOD, L_0 can not be measured at all by the standard test.

What is the DEPLETION of oxygen as a batch test progresses? Equation 2.5 defines the depletion as the initial DO value minus the DO value measured at any time t.

$$\text{Depletion} = [\text{DO}]_{\text{initial}} - [\text{DO}]_{\text{measured at time } t} \qquad (2.5)$$

If the measured DO values at any time t from Figure 25 are subtracted from the initial value, then Figure 26 results. Since the concentration of DO is proportional to the concentration of biodegradable material, then the BOD DEPLETION as a function of time will be comparable to the DO DEPLETION. The comparable function for BOD DEPLETION is shown in Figure 27. If y = the depletion of biodegradable material as a function of time, then,

$$y = L_0 - L \qquad (2.6)$$

and since from Equation 2.4,

$$L = L_0 \left[e^{-K_1 t} \right] \quad \text{or}$$
$$L = L_0 \left[10^{-kt} \right]$$

then the right hand side of Equations 2.4, can be substituted into Equation 2.6 for L, the remaining biodegradable material. Again, illustrating the procedure in both base e and base 10,

$$y = L_0 - L = L_0 - L_0 \left[e^{-K_1 t} \right] \quad \text{or}$$
$$y = L_0 - L = L_0 - L_0 \left[10^{-kt} \right] \qquad (2.7)$$

Combining terms yields the following expression for the depletion of remaining biodegradable material, called EXERTED BOD,

$$y = L_0 \left[1 - e^{-K_1 t} \right] \quad \text{or}$$
$$y = L_0 \left[1 - 10^{-kt} \right] \qquad (2.8)$$

It is very important to note that L_0 from Equations 2.3 or Equations 2.4, defined then as the initial concentration of biodegradable material, is the same value as L_0 in Equation 2.8. Here in Equation 2.8, however, L_0 is referred to as 'the ultimate BOD', and it corresponds to the upper plateau of Figure 27. You should realize that the value obtained for

48

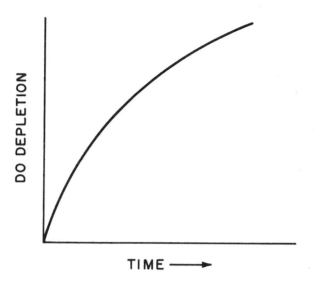

FIGURE 26. DO Depletion with Time.

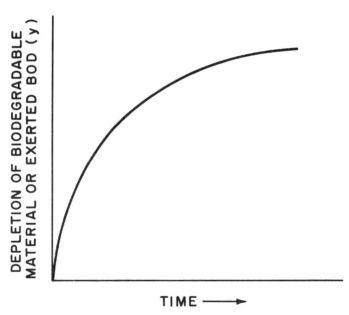

FIGURE 27. BOD Depletion with Time.

the ultimate depletion of BOD equals the initial value of the remaining BOD.

The 5-day BOD test measures EXERTED BOD. It is a standard test, carried out anywhere in the world according to the following procedure. The actual BOD determination is carried out in a special bottle called a BOD bottle. The volume of the bottle is 300 ml. A measured volume of wastewater is introduced into the bottle and the remaining volume filled with dilution water which is saturated with dissolved oxygen. The dilution water is prepared prior to the beginning of the test by aerating tap water with compressed air for 24 hours. The concentration of DO in the dilution water then approximates the saturation concentration of about 9 mg/ℓ. After the bottle is completely filled with the appropriate volumes of wastewater sample to be analyzed and dilution water, the initial DO is measured, the bottle is stoppered so that no oxygen can leak in or out, and the bottle is incubated at 20° C. for 5 days. During that five days, microorganisms in the wastewater will utilize dissolved oxygen from the dilution water and biodegradable material from the wastewater to carry out their aerobic metabolism according to Equation 1.3. The biochemical reaction will result in a depletion of biodegradable material in the wastewater and a depletion of dissolved oxygen from the dilution water. In order to measure the DO depletion accurately, the test bottle should have been set up so that the DO depletion will be greater than 2 mg/ℓ and the residual DO greater than 1 mg/ℓ. Therefore, the problem is to introduce the proper wastewater sample size on Day 1, so that on Day 5 the DO depletion will have fallen within this range. The test, therefore, must be carried out in a wide range of wastewater sample sizes in order to bracket the true value, or else one must have a prior estimate of the BOD to be determined in order to reduce the range of wastewater sample sizes.

From the description of the standard BOD test, you could probably deduce that the standard test can determine only one value, i.e., the depletion of biodegradable material occurring in the 5 day test period. The standard test does not determine a whole continuum of values illustrated in Figure 27. In addition, the standard test does not measure L_0 either as the initial concentration of remaining biodegradable material or the ultimate depletion of biodegradable material. This value of L_0 can only be measured by modifying the standard BOD test in order to allow it to continue incubating for an 'infinite' time period until no further oxygen depletion occurs.

There are two types of of BOD tests:

1. Without 'seed' (inherent microorganisms used)

2. With 'seed' (microorganisms added)

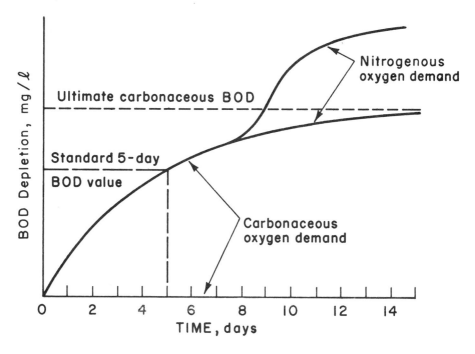

FIGURE 28. Carbonaceous and Nitrogenous BOD Exerted Stepwise.

Without seed: A wastewater to be tested will usually contain enough acclimated microorganisms already in the sample to carry out the necessary biodegradation. Figure 28 illustrates a result from a test with acclimated carbonaceous bacteria, but the nitrogenous bacteria exhibit a time lag.

The appropriate equation used for calculating BOD in this case is Equation 2.9.

$$\text{EXERTED BOD} = y_5 = \frac{\left[[DO]_i - [DO]_f\right]\left[\text{bottle volume}\right]}{[\text{sample size}]} \qquad (2.9)$$

where the subscripts i and f indicate initial and final values, respectively, and y_5 is the exerted BOD measured in 5 days.

With seed: Microorganisms must be provided for the sample by adding a seed organism, settled out from a wastewater, for example. If this seed is acclimated to the wastewater being tested, then again, no time lag will be observed. The graphical results of this test are the same as Figure 28, but the appropriate equation for calculating the BOD is modified from Equation 2.9.

$$y_5 = \left\{ \left[[DO]_i - [DO]_f\right]_w - \left(\left[[DO]_i - [DO]_f\right]_{bl} (fr) \right) \right\} \frac{[\text{bottle volume}]}{[\text{sample size}]} \qquad (2.10)$$

51

where, the subscripts w and bl stand for the wastewater sample and a blank, respectively.

A blank must be run simultaneously along with the test bottle of wastewater. The blank contains all ingredients except the wastewater itself. The fr term in Equation 2.10 is the fraction of seed in the wastewater sample per fraction of seed in the blank.

Limitations of the BOD test are as follows:

1. The wastewater sample size must be optimum in order to deplete the dissolved oxygen in the bottle more than 2 mg/ℓ and not more than the amount to retain a 1 mg/ℓ residual DO in the bottle.

2. The temperature must remain at 20° C.

3. No air may leak into the bottle.

4. The test must be carried out for 5 days.

EXAMPLE 1. Calculate the 5-day exerted BOD if $k = 0.1$ day^{-1} and $L_0 = 290$ mg/ℓ.

SOLUTION:

$$y = L_0 \ \ [1 - 10^{-kt}]$$

$$y = 290 \text{ mg/}\ell \ \ [1 - 10^{-(0.1 \text{ day}^{-1})(5 \text{ day})}]$$

$$y = 290 \text{ mg/}\ell \ \ [1 - 0.32]$$

$$y = 198 \text{ mg/}\ell$$

EXAMPLE 2. Calculate the 5-day exerted BOD from a test in which the following DO measurements have been determined for 10.0 ml of water sample used in the BOD bottle. (fr = 0.6 for this seeded test) :

for the blank:

$$DO_{initial} = 9.1 \text{ mg/}\ell$$

$$DO_{final} = 8.0 \text{ mg/}\ell$$

for the wastewater:

$$DO_{initial} = 8.0 \text{mg/}\ell$$

$$DO_{final} = 3.4 \text{mg/}\ell$$

SOLUTION: Using Equation 2.10,

$$y = \left\{ \left(8.0 \, \tfrac{mg}{\ell} - 3.4 \, \tfrac{mg}{\ell} \right) - \left[\left(9.1 \tfrac{mg}{\ell} - 8.0 \, \tfrac{mg}{\ell} \right)(0.6) \right] \right\} \left\{ \tfrac{300 \ ml}{10.0 \ ml} \right\}$$

$$= \{ (4.6 \ mg/\ell) - (0.7 \ mg/\ell) \} \ 30$$

$$= 117.0 \ mg/\ell$$

Other analytical techniques developed to eliminate disadvantages of the BOD test are not as diagnostic as the BOD test. The following tests do not imitate natural processes in the BOD bottle, in the natural environment, or in a biological wastewater treatment plant since they do not measure the BIOCHEMICAL oxygen demand. These rapid techniques have been developed to overcome the 5-day lag period required for the BOD test. They are merely indicators of BOD values, and, when these tests are carried out continuously or in succession, they can indicate changes in BOD concentrations with time. However, no *a priori* correlation exists between the BOD test and any of the following methods. In order to correlate any of the following tests with the standard BOD test, both the BOD test and the test to be correlated must be carried out on the same number of statistically significant samples.

1. COD or **Chemical Oxygen Demand**: This analysis utilizes the oxidizing agent dichromate, added to the wastewater sample in order to oxidize the pollutants which can be CHEMICALLY oxidized. The reaction taking place is as follows:

$$\text{organic matter} \ + \ Cr_2O_7^{-2} \ \rightarrow \ CO_2 \ + \ H_2O \ + \ Cr_2O_4^{-2}$$

| reduced reactant | oxidizing agent | oxidized products | reduced oxidizing agent |

The sample to be measured and the dichromate oxidizing agent are mixed and heated for 1 hour in a flask attached to a reflux condenser. After an hour's time, the above reaction is completed, and the flask is cooled in order to measure the results. As indicated in the chemical reaction above, the amount of oxidizing agent used up to convert the organic matter to CO_2 is directly proportional to the oxygen demand of the wastewater. The difficulty in this case is to measure the amount of oxidizing agent used up in the above reaction once the reaction is complete. To overcome this difficulty, a blank is run (containing all the reagents, but eliminating the wastewater sample to be analyzed) in order to determine the amount of oxidizing agent which had been added initially to the above reaction. Since the blank contains no organic matter, no oxidation reaction will occur in the blank during the one hour heating period. The amount of oxidizing agent initially added to all the samples will remain in the blank after the test is complete. The concentration of oxidizing agent in the blank therefore, represents the initial concentration of oxidizing agent added to all the samples for the particular test run. The organic matter present in the wastewater samples, on the other hand, will be oxidized during the heating period and the proportional amount of oxidizing agent will be reduced. The amount of oxidizing agent used up is represented by the following expression:

$$\text{Ox Agt}_{\text{used up}} = \text{Ox Agt}_{\text{remain'g bl}} - \text{Ox Agt}_{\text{remain'g ww}}$$

Both the wastewater samples and the blank must then be titrated with a standard reducing agent in order to determine the amount of remaining oxidizing agent in the respective flasks. The standard reducing agent used is ferrous ammonium sulfate, and during the titration, the ferrous ion in the titrant is oxidized to ferric ion ($Fe^{+2} \rightarrow Fe^{+3}$). The titration reaction used to determine this value is as follows:

$$Cr_2O_7^{-2} \quad + \quad Fe^{+2} \quad \rightarrow \quad Fe^{+3} \quad + \quad Cr_2O_4^{-2}$$

unused oxidizing agent	standard reducing agent	oxidized reducing agent	reduced oxidizing agent

An example calculation follows:

EXAMPLE 3. A ferrous ammonium sulfate titrant is standardized against 5.0 ml of known 0.25 N standard potassium dichromate solution. In the standardization titration, 22.30 mls of titrant are used to reach a neutralization point. The titrant is then used in a COD analysis of 5.0 ml of wastewater. The difference in titrant volume between the blank and the sample is 5.16 ml. Calculate the titrant normality and the COD concentration in the wastewater.

SOLUTION:

$$\text{Titrant Normality} = \frac{(V_{\text{standard}})(N_{\text{standard}})}{V_{\text{titrant}}}$$

54

$$N_{\text{titrant}} = \frac{(5.0 \text{ ml})(0.25\text{N})}{(22.30 \text{ ml})}$$

$$N_{\text{titrant}} = 0.056\text{N}$$

NOTE: '0.056 N' means 0.056 'equivalents per liter'

$$\text{mg}/\ell \text{ COD} = \frac{\left[[V_{bl} - V_{s}]_{\text{titrant}}[N]_{\text{titrant}}\right]}{[\text{Sample Size}]} \frac{\left[\text{Equiv Wt}_{\text{oxygen}}\right]}{[\text{equivalent}]}$$

where the subscripts bl and s indicate the blank and sample, respectively,

$$\text{mg}/\ell \text{ COD} = \frac{(5.16 \text{ ml})(0.056 \text{ N})}{(5.0 \text{ ml})} \frac{(8000 \text{ mg})}{(\text{equivalent})}$$

$$\text{COD} = 462 \text{ mg}/\ell$$

2. TOD or **Total Oxygen Demand**: This analysis is carried out in a combustion furnace where oxygen gas flows directly into a combustion chamber containing the wastewater sample. The reaction taking place is as follows:

$$\text{pollutants} + O_2 \rightarrow \text{oxides} + H_2O$$

The amount of O_2 used ($[O_2]_{\text{in}} - [O_2]_{\text{out}}$) can be measured by a gas thermal conductivity detector attached downstream from the furnace. The gas flows from the furnace into the detector, and this type of detector measures the difference in electrical potential of the furnace exit gases when just the background oxygen is exiting and when the sample oxidation products are exiting. The height of the plotted potential difference peak is proportional to the concentration of oxidation products, which in turn, are proportional to the oxygen demand of the wastewater. In order to correlate the detector analogue output with digital concentration values, a calibration curve must be determined using a known concentration of standard. The unknown wastewater sample detector results can then be compared with the calibration curve of the known standard. Figure 29 illustrates both the calibration curve and the results of an unknown synthethic wastewater sample of glucose. A sample calculation for this technique follows:

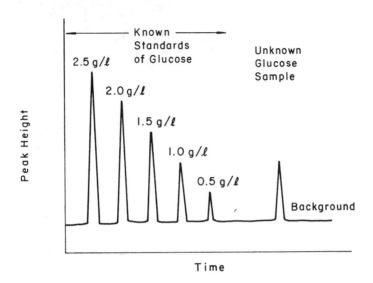

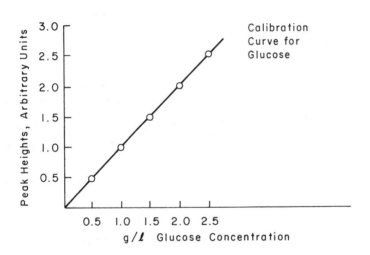

FIGURE 29. Calibration Analyses and Sample Analysis
for a Thermal Conductivity Detector.

EXAMPLE 4. Figure 29 indicates that the concentration of glucose in the unknown wastewater sample is 1 gm/ℓ. Calculate the resulting TOD.

SOLUTION: The following reaction applies for the oxidation of glucose:

$$C_6H_{12}O_6 + 6\,O_2 \rightarrow 6\,CO_2 + 6\,H_2O$$

Mol Wts	180	32	44	18

$$mg/\ell \; TOD = \frac{\left[1 \; gm \; glu\right]}{\left[\ell\right]} \frac{\left[mole \; glu\right]}{\left[180 \; gm\right]} \frac{\left[6 \; mole \; O_2\right]}{\left[mole \; glu\right]} \frac{\left[32 \; gm \; O_2\right]}{\left[mole \; oxygen\right]} \frac{\left[1000 \; mg\right]}{\left[gm\right]}$$

$$TOD = 1067 \; mg/\ell$$

---

3. TOC or **Total Organic Carbon**: This analysis also is carried out in a combustion furnace using pure oxygen gas passing over the wastewater sample. Here, however, the CO_2 produced is measured by an infrared spectrophotometer. This detector measures the difference in absorption of infrared light at a specific light frequency for CO_2 when a wastewater sample is absent and when a wastewater sample is oxidized in the furnace to produce CO_2. The plotted output and calibration for this detector are comparable to that shown in Figure 29, and the following example indicates the similarity to the TOD calculation. In this case, the reaction taking place is as follows:

$$organic \; carbon + O_2 \rightarrow CO_2 + H_2O$$

EXAMPLE 5. Calculate the TOC for the glucose solution in Figure 29.

SOLUTION: The same chemical reaction applies as in Example 4, but the calculation differs slightly.

$$mg/\ell \; TOC = \frac{\left[1 \; gm \; glu\right]}{\left[\ell\right]} \frac{\left[mole \; glu\right]}{\left[180 \; gm\right]} \frac{\left[6 \; mole \; carbon\right]}{\left[mole \; glu\right]} \frac{\left[12 \; gm\right]}{\left[mole \; carbon\right]} \frac{\left[1000 \; mg\right]}{\left[gm\right]}$$

$$TOC = 400 \; mg/\ell$$

In a treatment plant, then, with these three rapid tests, an operator can determine whether or not a reactor is presently working. These tests—COD, TOD, and TOC—are merely quick indicators of BOD. It can be repeated that, independently, none of them gives any accurate measure of BOD. If, however, many different samples are tested simultaneously for BOD and any of the other tests above, a statistical correlation of the quick test to the BOD test can be calculated. In addition, any of these tests can be used to determine relative changes in BOD concentrations. For example, any of these tests may be used to quickly determine relative incoming and outgoing concentrations in a specific reactor or in an entire biological wastewater treatment plant.

B. Solids Analyses

Solid pollutants in wastewater can be classified as suspended or dissolved. A ratio of these two classes can indicate which type of treatment would be necessary to treat the wastewater. For example, if the wastewater contains a low concentration of suspended solids, but a high concentration of dissolved solids, then the engineer will know that removal of dissolved pollutants is paramount.

The Total Solids (TS) is a sum of the Suspended Solids (SS) and the Dissolved Solids (DS).

$$TS = SS + DS \qquad (2.11)$$

<u>SS</u>. Suspended solids are measured by filtering a measured volume of wastewater through a pre-weighed filter paper. The filtrate is discarded or used to measure some dissolved pollutant. The filter paper and solids are dried and re-weighed.

$$\text{mg}/\ell \;\; SS = \big[[\text{wt paper + solids}] - [\text{wt paper}]\big]\frac{[1000 \text{ ml}]}{[\text{sample size}]} \qquad (2.12)$$

<u>TS</u>. Total solids are measured by evaporating a known volume of wastewater in a pre-weighed evaporating dish. When the liquid has completely evaporated, the dish is re-weighed.

$$\text{mg}/\ell \;\; TS = \frac{[(\text{wt dish + solids}) - (\text{wt dish})][1000\text{ml}]}{\text{sample size}} \qquad (2.13)$$

<u>DS</u>. The Dissolved Solids are usually calculated by difference from Equation 2.11.

Table 5 indicates typical values for wastewaters which are strong, medium, and weak. This table contains the terms 'fixed' and 'volatile'. Fixed solids are usually inert, inorganic solids which are not combusted at 600° C., and volatile solids are usually organic solids which will combust at 600° C. The previous three types of solids may be further delineated into fixed and volatile categories, therefore, by subjecting the dried, weighed solids and the filter paper to combustion at 600° C. After combustion, only the fixed solids will remain to be weighed. The volatile solids will equal the weight difference of the solids before and after combustion. These various solids types are abbreviated as above and as in the table below.

TABLE 5

Typical Composition of Domestic Wastewaters
All values except settleable solids
are expressed in mg/ℓ

Constituent		Concentration		
		Strong	Medium	Weak
Solids, total	(TS)	1,200	700	350
Dissolved, total	(TDS)	850	500	250
Fixed	(FDS)	525	300	145
Volatile	(VDS)	325	200	105
Suspended, total	(TSS)	350	200	100
Fixed	(FSS)	75	50	30
Volatile	(VSS)	275	150	70
Settleable solids, (ml/liter)		20	10	5
Biochemical oxygen demand, 5 day, 20° C		300	200	100
Total organic carbon (TOC)		300	200	100
Chemical oxygen demand (COD)		1,000	500	250
Nitrogen, (Total as N)		85	40	20
Organic		35	15	8
Free ammonia		50	25	12
Nitrites		0	0	0
Nitrates		0	0	0
Phosphorus (Total as P)		20	10	6
Organic		5	3	2
Inorganic		15	7	4
Chlorides		100	50	30
Alkalinity (as CaCO3)		200	100	50
Grease		150	100	50

Examination of Table 5 indicates that almost all the parameters used to describe wastewater strengths now have been discussed. The alkalinity will be discussed in Chapter 3, but the grease parameter will not be discussed in detail.

And, finally, it should be noted that the incoming wastewater will contain all the characteristics present in the drinking water prior to its use. For example, if the drinking water provided is high in carbonates or chlorides, then the wastewater will contain high levels of these dissolved solids. In addition to those dissolved solids already present in the drinking water supplied, the used water will contain less desirable pollutants in the form of BOD and suspended solids.

This factor can not be ignored when wastewater is re-used directly or indirectly, as in a river basin. Engineers and citizens must analyze the system according to Figure 1 to determine the degree of treatment necessary for future use of the receiving water body and the resulting cost. Table 6 presents criteria needed to determine use categories. The engineer and planner must decide what use the water will serve and therefore what degree of treatment must be imposed on wastewater to be discharged to the system.

TABLE 6
Water Quality for Stream Uses

CLASS	USE	QUALITY CRITERIA
A	Potable water supply	Microbiological counts, color, turbidity, pH, Dissolved Oxygen, toxic materials, taste, odor, temperature.
B	Bathing, primary contact, recreation, fishing	Same as A but less stringent levels
C	Industrial, agricultural, fishing, navigation	Dissolved oxygen, pH, suspended solids, temperature
D	Cooling, navigation	Floating material, pH, suspended solids

REFERENCES

[1] R. M. Brown, et al., *Water and Sewage Works*, **117**, (10), 339 (1970).
[2] W. Stumm and J. J. Morgan, *Aquatic Chemistry*, Wiley Interscience, New York, 1970.
[3] C. N. Sawyer and P. L. McCarty, *Chemistry for Sanitary Engineers*, McGraw-Hill, New York, 1968.
[4] *Pub. Health Serv. Drinking Water Standards*, (PHS Publication No. 956), 1962.
[5] J. B. Carberry and M. W. Tenney, *J. Water Pollution Control Fed.*, **45**, 2444 (1973).

HOMEWORK PROBLEMS

1. List the advantages and disadvantages of the BOD test.

2. What is the advantage of the BOD test over the TOC, TOD, and COD tests?

3. What is the principal disadvantage of each of the TOC, TOD, and COD tests?

4. If a 1 gm/ℓ solution of glucose, which is totally biodegradable, yields a BOD of 1067 mg/ℓ, what will the TOD measure? the TOC? How biodegradable is the sample?

5. If a 5 ml sample of wastewater produced the following data in a BOD determination, what is the BOD exerted in 5 days?
$$DO_{initial} = 7.7 \text{ mg}/\ell$$
$$DO_{5days} = 2.8 \text{ mg}/\ell$$

6. Calculate the ultimate BOD, L_0, in Problem 5 if the biodegradation rate constant k (base 10) is 0.35 day^{-1}.

7. If a wastewater treatment plant achieved a 90% efficiency for the wastewater in Problem 5, what BOD concentration would be discharged to the environment?

8. Calculate the TOD and TOC of a solution containing 750 mg/ℓ of glucose.

9. Calculate the COD for a glucose sample using the following data:

    sample size = 10.0 ml
    normality titrant = 0.080 N
    volume titrant for blank = 22.84 ml
    volume titrant for sample = 21.60 ml

10. Wastewater with 4000 mg/ℓ glucose ($C_6H_{12}O_6$) from a candy factory and 300 mg/ℓ of fertilizer with equal concentrations of nitrogen and phosphorus must be discharged to a biological wastewater treatment plant. Use the molecular formula for a microorganism (on p. 16) to determine whether carbon, nitrogen, or phosphorus is the limiting reactant for the biodegradation of the wastewater.

11. Sample sizes of 1.0 ml, 5.0 ml, 10.0ml and 15.0 ml were used to determine the BOD of a wastewater. The 5-day BOD was 150 mg/ℓ for each sample size and the initial DO measured for each determination was 8.2 mg/ℓ. Find the final DO for each sample size and indicate whether each determination for the various sample sizes is valid. Which sample size should be used to analyze this wastewater?

COMPUTER PROBLEM
Determine the sample size limits for a standard BOD test on the wastewater described in Problem 5. HINT: The DO depletion in the BOD bottle must be at least 2 mg/ℓ, and the final DO must be at least 1 mg/ℓ. Assume that the initial DO measurement will not change significantly with the small range of acceptable sample small range of acceptable sample sizes of wastewater used.

Chapter 3

SOLUTION AND SURFACE CHEMISTRY

A. Homogeneous and Heterogeneous Aqueous Concepts

Water is the basis of living creatures on this planet. Living cells are over 90% water, and fundamental biological functions occur inside the cell within the liquid medium. Some functions inside the living cell occur in the homogeneous liquid phase with dissolved reactants and products. Other reactions occur utilizing heterogeneous 'solid' structures as mediators inside the cells; some reactions may produce heterogeneous, insoluble products. For our purpose, we need not be concerned with fine detail, so it is helpful to classify our systems as homogeneous or heterogeneous, and look at the chemistry of both classes. The first class deals with solutions of substances DISSOLVED in suspending medium, usually water. The heterogeneous case deals with solid particles SUSPENDED in a liquid, again, usually in water suspensions.

The first section of this chapter will examine solution chemistry, particularly water solutions, and this will be primarily a review of freshman chemistry concepts. These familiar concepts will then be applied to environmental engineering problems. The second section will present some new concepts concerning surface effects on solid particulates suspended in water. Here we will see how the surface properties of these solids can be modified to enhance favorable engineering separation processes.

B. Solution Chemistry

Water is composed of molecules of H_2O, which at standard temperature and pressure exist in the liquid state. These bent molecules can dissociate into charged ions according to the following water dissociation reaction:

$$H\text{—}O\text{—}H \rightarrow OH^- + H^+$$

The tendency for this dissociation to occur spontaneously is very slight and the degree of dissociation is an index of this tendency, called the water dissociation constant, K_W. For the water dissociation reaction above, the **dissociation constant** is expressed as follows:

$$K_W = [OH^-][H^+] = 1 \times 10^{-14} \text{ moles}^2/\ell^2 \tag{3.1}$$

If we take logs of both sides of Equation 3.1, the following expression results:

$$\log K_W = \log\left[[H^+][OH^-]\right] = -14$$

Under neutral conditions, $[H^+] = [OH^-]$ and therefore,

$$\log[H^+] = \log[OH^-] = -7$$

And, if pH is defined as the negative log of $[H^+]$, then, under neutral conditions, pH $= 7$. The same methodology can be used to define pOH as the negative log of $[OH^-]$ and pK_W as the negative log of K_W. At **neutrality**, then, pOH $= 7$, also. Under all conditions of neutrality or non-neutrality, however, Equation 3.1 controls the system. For example, if $[H^+] = 1 \times 10^{-4}$ moles/ℓ, then pH $= 4$; and from Equation 3.1, $[OH^-] = 1 \times 10^{-10}$ moles/ℓ with pOH $= 10$. In this case, the system is **acidic**. If $[H^+]$ is decreased to 1×10^{-9} moles/ℓ, then the pH $= 9$ and the conditions of the system would be **basic**. Since pH is a negative log value, the pH decreases with increasing $[H^+]$ and the system is acidic at pH < 7. Under basic conditions, pH > 7, when $[H^+]$ decreases to some concentration less than 1×10^{-7} moles/ℓ. Table 7 illustrates increasing hydrogen ion concentrations and comparable decreasing pH. These calculations and conversions should become routine.

In addition to pH and pOH, as mentioned previously, the negative log approach can also be used to specify pK_W as $-\log K_W$. And pK_W, then, equals 14. This 'p-notation' is a very convenient way to express

64

Table 7

[H$^+$] Concentrations and pH

pH	[H$^+$]
6	?
5.2	?
5.0	1×10^{-5}
4.2	6.3×10^{-5}
4.0	1×10^{-4}
3.5	3.2×10^{-4}
?	1×10^{-3}
?	7.4×10^{-2}

and calculate any of these values [1]. The pH is the most important parameter of a system and controls many factors. For example, living cells can exist within the pH range $5 < $ pH $ < 9$; certain compounds are insoluble at high pH's and will precipitate out when basic conditions prevail; and, finally, suspended particles in a wastewater are strongly influenced by pH conditions. The remaining sections of this chapter will present details of changing pH mechanisms and how they effect both natural and engineered systems.

1. Acid-Base Dissociation

Acids are compounds usually composed of a positively charged hydrogen ion, H$^+$, and a negatively charged anion. (Anions are negatively charged, and, under the influence of an electrical potential, they migrate toward the positively-charged anode.) Such an anion commonly combining to form an acid is chloride ion, Cl$^-$, which forms hydrochloric acid when combined with H$^+$. Chloride ion possesses one negative charge, so one chloride ion will combine with one hydrogen ion to form one molecule of hydrochloric acid, HCl. Such an acid is classified as mono-protic, since it contains only one hydrogen ion. (One hydrogen ion is one proton.) Di- and tri-protic acids are formed from poly-charged anions such as sulfate, SO_4^{-2}, or borate, BO_3^{-3}. Each hydrogen ion in these poly-protic acids dissociates from its associated anion in a stepwise sequence. And the tendency for each stepwise dissociation can be measured by its respective stepwise dissociation constant, K_{A_1}, K_{A_2}, etc. For example, boric acid, the tri-protic acid of borate ion will dissociate by the following sequence of reactions [2].

65

$$\begin{array}{lcc}
 & K_A & pK_A \\
H_3BO_3 \rightarrow H^+ + H_2BO_3^- & 5.8 \times 10^{-10} & 9.24 \\
H_2BO_3^- \rightarrow H^+ + HBO_3^{-2} & 1.8 \times 10^{-13} & 12.7 \\
HBO_3^{-2} \rightarrow H^+ + BO_3^{-3} & 1.6 \times 10^{-14} & 13.8
\end{array} \quad (3.2)$$

As can be seen from the dissociation constants in the above sequential dissociation progression, each stepwise dissociation is less and less likely to occur. Moreover, since each sequential dissociation produces a hydrogen ion, the prior presence of hydrogen from a strong acid will shift the direction of the above dissociations to the left, according to Le Chatlier's Principle.

Such acids with small dissociation constants are considered to be weak acids, and their dissociations are controlled by the prevailing pH. Strong acids, on the other hand, are virtually 100% dissociated and control the pH of the system. If, for example, a large volume of a strong acid such as HCl at a high concentration is discharged into a small stream or lake, the pH might be depressed below the life-limiting value of pH = 5, and all living forms in the system would die. Many ecologists have claimed such a result has occurred due to acid rain in the Adirondack Mountains of New York State.

2. Acid-Base Neutralization

Strong bases dissociate almost 100%, just like strong acids. Examples of strong bases are sodium hydroxide, NaOH, and potassium hydroxide, KOH. The dissociation reaction of such a strong base is as follows:

$$NaOH \rightarrow Na^+ + OH^-$$

The equilibrium position of this dissociation lies almost completely to the right, as this strong base is virtually 100% dissociated. Weak bases, on the other hand, dissociate to a lesser degree, and the tendency for them to dissociate is measured by their dissociation constant, K_B. A common weak base is ammonium hydroxide, NH_4OH. Its dissociation reaction, with its K_B and pK_B are illustrated below:

$$NH_4OH \rightarrow NH_4^+ + OH^-$$

$$K_B = 1.8 \times 10^{-5} \quad \text{and} \quad pK_B = 4.74$$

66

Since acids dissociate to produce hydrogen ion, and bases dissociate to produce hydroxyl ion, these two dissociations together produce the same two ions as the water dissociation reaction. Acids and bases can be combined, then, to form water molecules, and this reaction is called a neutralization reaction. The neutralization reaction is the reverse of the water dissociation reaction shown at the beginning of this chapter. Quite often, an unknown concentration of an acid or base is titrated by a known concentration and measured volume of the other. If, for example, a strong acid is titrated by a strong base, the end point determination is easy and often carried out by automatic instrumentation. An illustration of such a titration is shown in Figure 30 [3]. Here, the 1:1 stoichiometric reaction of hydrogen and hydroxyl ions produces a straightforward curve which is easily interpreted. A measured volume of acid with unknown concentration is titrated by a measured volume of base with known concentration. After each volumetric addition of base, the pH is measured and plotted, as in Figure 30. At first, each small addition of strong base is easily neutralized by the prevailing hydrogen ion, and the pH does not change significantly, due to the excess hydrogen ion remaining untitrated. As the titration approaches the neutralization point, however, the logarithmic value of concentrations changes dramatically, and it is usually easy to find the inflection point of the titration curve. It should be pointed out that many automatic titrators no longer provide such an analog output on a routine basis, but instead, measure titrant volumes required to reach a pre-set digital signal endpoint. Some can be programmed to calculate differentials in order to determine the inflection point of the titration curve, and some can be programmed to calculate concentrations in the unknown samples. These instruments are often installed on line in some processes to sample the process stream continuously or intermittantly and signal an alarm to call for manual adjustment, or trigger an automatic adjustment if the system is electronically- or computer-controlled.

Weak acids may be titrated with strong bases, and, conversely, weak bases may be titrated with strong acids. It is these cases which most interest environmental engineers. An illustration of such titrations is shown in Figure 31. This figure illustrates the titration of three weak acids by a strong base, NaOH [3]. Here, the equivalence point is not necessarily at pH $= 7$, and the titration curve is not as distinct and easy to interpret as that for strong acids shown in Figure 30. The reason for this lack of distinct inflection is due partly to the character of the acid itself and partly to its concentration.

For the development of useful concepts of weak acids and bases, it

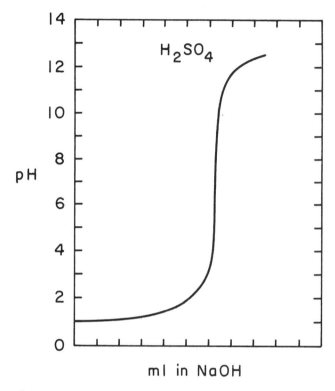

FIGURE 30. Strong Acid Titration by a Strong Base.

might be easiest to begin by using an example and then draw generalizations from this case. For this purpose, we can use the weak acid, acetic acid. This acid is an organic acid, common in many products such as vinegar, and its titration curve is shown in Figure 31. A shorthand notation will be used, for the associated acid, HAc, and for the dissociated acetate anion, Ac^-. The dissociation reaction and corresponding dissociation constant expression are shown below:

$$HAc \rightarrow H^+ + Ac^-$$

$$K_A = \frac{[H^+][Ac^-]}{[HAc]} = 1.8 \times 10^{-5} \text{ moles}/\ell$$

Before any titration of this acid begins, only these three components are present in this solution. From the dissociation reaction, it can be seen that for every mole of acid dissociated, one mole of hydrogen ion and one mole of acetate ion are produced. Therefore, before the titration begins, $[H^+] = [Ac^-]$, and $[H^+]$ can be substituted into the

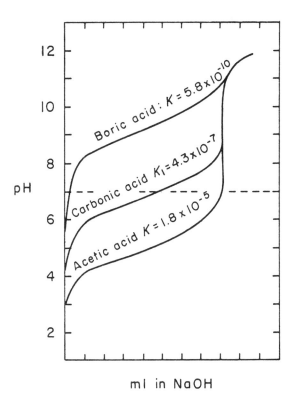

FIGURE 31. Titration of Weak Acids by Strong Base.

dissociation constant expression for [Ac⁻]. The following expression results:

$$K_A = \frac{[H^+]^2}{[HAc]} \qquad \text{and} \qquad [H^+] = \left[K_A[HAc]\right]^{\frac{1}{2}}$$

It is helpful to consider that all the acid is either associated or dissociated. If A is the total acid, associated and dissociated, then this concentration may be calculated by the following expression:

$$[A] = [HAc] + [Ac^-] \tag{3.3}$$

From Equation 3.3, $[HAc] = [A] - [Ac^-]$, and therefore,

$$[H^+] = \left[K_A([A] - [Ac^-])\right]^{\frac{1}{2}}$$

And since the total acid concentration, [A] is much greater than [Ac⁻], then, for a weak acid, [Ac⁻] can be eliminated from the equation above, yielding,

69

$$[H^+] = \left[K_A[A]\right]^{\frac{1}{2}}$$

Taking negative logs of both sides of this expression yields a very useful formula for calculating the initial concentration of acid present:

$$-\log[H^+] = -\frac{1}{2}\log K_A - \frac{1}{2}\log[A] \qquad \text{or in} \quad p - \text{notation}$$

$$pH = \frac{1}{2}pK_A + \frac{1}{2}p[A] \tag{3.4}$$

$$pH = \frac{1}{2}\left(pK_A + p[A]\right)$$

Therefore, from Figure 31, we can calculate the initial concentration of acid present. From the titration curve in Figure 31, the initial pH = 2.98, and the dissociation constant for acetic acid is given as $K_A = 1.8 \times 10^{-5}$. Therefore, from Equation 3.4,

$$2.98 = \frac{1}{2}\left(4.74 + p\,[A]\right) \quad \text{and}$$

$$p\,[A] = 1.22$$

Therefore, the initial $[A] = 0.06$, or 6×10^{-2} moles/ℓ.

When the titration is begun, strong base is added to the solution, and the strong base dissociates almost 100% into the basic cation, B^+, and the hydroxyl ion, OH^-, by the following reaction:

$$\text{'}\ BOH \rightarrow B^+ + OH^-$$

and the hydroxyl ions react with the dissociated hydrogen ions from the acid to carry out the reverse reaction of Equation 3.1

$$OH^- + H^+ \rightarrow H_2O.$$

The concentration of hydrogen decreases, therefore, and the associated acid molecule tends to dissociate more and more as the hydrogen ions are neutralized. Consequently, [HAc] decreases and $[Ac^-]$ increases as the titration progresses. Also, $[B^+]$ increases and $[H^+]$ decreases so that electroneutrality is preserved as the pH increases.

At the midpoint of the titration, when 50% of the acid has been, neutralized, $[B^+] = \frac{1}{2}[A]$. Since $[A] = [HAc] + [Ac^-]$, at the midpoint of the titration the associated and dissociated forms of acetate ion

70

will be equal, i.e., $[\text{HAc}] = [\text{Ac}^-]$. And furthermore, from the equalities just presented, the following is true at the midpoint of the titration:

$$[\text{B}^+] = \frac{1}{2}[\text{A}] = [\text{Ac}^-] = [\text{HAc}]$$

therefore, at the titration midpoint, the dissociation expression becomes

$$K_A = \frac{[\text{H}^+](\frac{1}{2}[\text{A}])}{\frac{1}{2}[\text{A}]} \quad \text{or} \quad \text{p}K_A = \text{pH} \tag{3.5}$$

When 50% of the acid is neutralized, then, the pH changes little with large volumes of titrant added, because as soon as the acid dissociates, the hydrogen ion produced is neutralized by the hydroxyl ion from the added base titrant. The inflection point of the curve betrween the onset and end of the titration marks the midpoint of the titration and hence the pK_A. After the midpoint, the pH increases gradually then, as shown in Figure 31.

At the endpoint of the titration, when all the acid has been neutralized, the pH increases dramatically due to the continually increasing concentration of unreacted excess OH^- added to the solution. At the endpoint of the titration, $[\text{B}^+] = [\text{A}]$, and $[\text{H}^+] = 0$. To conserve electroneutrality,

$$[\text{B}^+] = [\text{Ac}^-] + [\text{OH}^-]$$

and, therefore, since $[\text{B}^+] = [\text{A}]$,

$$[\text{A}] = [\text{Ac}^-] + [\text{OH}^-].$$

And, since by definition,

$$[\text{A}] = [\text{Ac}^-] + [\text{HAc}]$$

then,

$$[\text{HAc}] = [\text{OH}^-]$$

and from Equation 3.1,

$$[\text{OH}^-] = K_W/[\text{H}^+]$$

therefore the following expression results:

71

$$K_A = \frac{[\text{H}^+][\text{Ac}^-]}{[\text{HAc}]}$$

$$= \frac{[\text{H}^+]]\text{Ac}^-]}{[\text{OH}^-]}$$

$$= \frac{[\text{H}^+][\text{Ac}^-]}{K_W/[\text{H}^+]}.$$

Therefore,

$$\frac{1}{[\text{H}^+]^2} = \frac{[\text{Ac}^-]}{K_W K_A}.$$

Since, at the end of the titration, $[\text{Ac}^-] = [\text{A}]$,

$$\frac{1}{[\text{H}^+]^2} = \frac{[\text{A}]}{K_W K_A}.$$

Taking logs of this expression and converting to p-notation,

$$\text{pH} = \frac{1}{2} \left(\log \text{A} + \text{p}K_W + \text{p}K_A. \right) \tag{3.6}$$

And, continuing use of the previous example to illustrate the calculation of endpoint pH from Equation 3.6,

$$\text{pH} = \frac{1}{2}(-1.22 + 14 + 4.74) = \frac{1}{2}(17.52) = 8.76$$

Equation 3.6 indicates that both the character of the acid $(\text{p}K_A)$ and the concentration of the acid $(\log [\text{A}])$ determine the pH at the equivalence point.

A comparable case exists for weak bases when titrated with a strong acid. For this case, the titration curve is inverted, with a high initial pH before any strong acid is added and a low pH at the equivalence point. Also, one should note that weak acids are never titrated with a weak base, and visa versa, because the ion concentration would be so dilute.

3. Carbonate Equilibria and Natural Alkalinity

Poly-protic acids may be titrated and yield a pK_A and an equivalence point for each hydrogen ion in the molecule. The most important example in this case is carbonic acid. Carbonic acid is a weak di-protic acid which is introduced into natural waters by the weathering action of natural limestone strata in the earth's crust. To a lesser, but still significant extent, atmospheric CO_2 may become dissolved into natural surface waters and combine chemically with water molecules to form carbonic acid via the following reaction:

$$CO_2 + H_2O \rightarrow H_2CO_3$$

The limestone rock weathering reaction is as follows:

$$CaCO_3 + 2\,H_2O \rightarrow Ca(OH)_2 + H_2CO_3$$

lime-stone	calcium hydroxide	carbonic acid

These reactions are rather simplistic, but they serve to illustrate how both surface and ground waters acquire H_2CO_3, and how both a weak acid and a strong base are introduced to the natural water system. As both compounds ionize, the hydroxyl ion neutralizes the weakly dissociated hydrogen ion. In fact, the first titration curve for carbonic acid is shown in Figure 31. That curve represents the following dissociation reaction, followed by the neutralization reaction:

$$H_2CO_3 \rightarrow HCO_3^- + H^+$$

carbonic acid	bicarbonate ion	hydrogen ion

and, for this this dissociation reaction, the dissociation constant is expressed as follows:

$$K_{A_1} = \frac{[H^+][HCO_3^-]}{[H_2CO_3]}$$

Here, the second level subscript indicates the first dissociation reaction. From Figure 31, $K_{A_1} = 4.3 \times 10^{-7}$, and therefore, $pK_{A_1} = 6.37$. The neutralization of this first hydrogen produced from dissociation occurs in natural water systems, as there is an excess hydroxyl ion from calcium hydroxide produced from the weathering reaction. The

73

bicarbonate ion produced from the dissociation is comparable to the previous acetate case. Here, though, the bicarbonate ion contains an additional ionizable hydrogen ion. The second hydrogen ion dissociates in the second dissociation reaction, as follows:

$$HCO_3^- \rightarrow CO_3^{-2} + H^+$$

bicarbonate   carbonate
ion   ion

and

$$K_{A_2} = \frac{[H^+][CO_3^{-2}]}{[HCO_3^-]} = 4.7 \times 10^{-11}$$

And, therefore, $pK_{A_2} = 10.33$. By using the titration curve shown in Figure 31 along with Equation 3.4, the initial concentration of acid present in the sample may be calculated.

$$pH_{initial} = \frac{1}{2}(pK_{A_1} + p[A])$$

$$4.25 = \frac{1}{2}(6.37 + p[A])$$

$$p[A] = 2.13, \text{ and } [A] = 0.01 \text{ or } 1 \times 10^{-2} \text{ moles}/\ell$$

This initial concentration of acid, then, is neutralized in nature by the co-produced strong base, $Ca(OH)_2$. As the titration continues, $[H_2CO_3]$ decreases and $[HCO_3^-]$ increases. At pH = 6.37, $[H_2CO_3]$ = $[HCO_3^-]$. Figure 31 indicates that the first endpoint for carbonic acid is at a pH of about 8.5, where the sharp rise in the titration curve occurs. In practice, when, in fact, it is difficult to determine first and second dissociation constants and endpoints, it is often helpful to estimate the endpoint using Equation 3.7.

$$pH_{first\ endpoint} = \frac{1}{2}(pK_{A_1} + pK_{A_2}) \qquad (3.7)$$

For carbonic acid, $pK_{A_1} = 6.37$, and $pK_{A_2} = 10.33$. The above formula, then, produces a pH value of 8.35 for the first equivalence point of the titration shown in Figure 31. At this pH, $[H_2CO_3] = 0$, and with

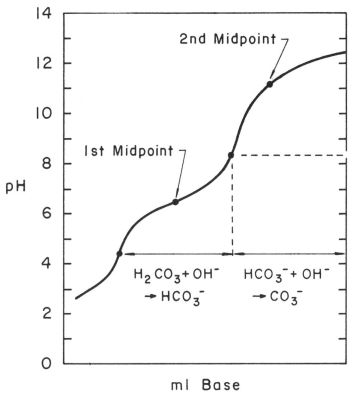

FIGURE 32. Carbonic Acid Titration.

increasing pH, bicarbonate ion begins to be converted to carbonate ion. Figure 32 shows both stepwise reactions in succession [3].

Table 8 summarizes this stepwise dissociation and progressive neutralization of carbonic acid with increasing pH. From this summary, it is apparent that at near-neutral pH, the bicarbonate ion is the dominant form. And, with a weak acid, the pH controls the dominant form. This carbonic acid/bicarbonate/carbonate system is very prevalent in natural systems due to the prevalence of limestone rocks in the earth's crust and the weathering process. The concentration of these forms of carbonic acid/bicarbonate/carbonate in natural waters is called the carbonate alkalinity concentration, and this carbonate alkalinity concentration is a very important characteristic of the water. This term 'alkalinity' is unfortunate, because it implies that the natural water is alkaline or basic due to the presence of these forms. This is definitely not the case, for the carbonate alkalinity actually provides a buffering capacity for the natural water. Figure 33 indicates that at a neutral pH which would be desirable in a natural water system, the dominant ion

75

Table 8

Summary of Titration Reactions

pH	dominant ion/molecule	reaction
1- 6.37	H_2CO_3	$H_2CO_3 + OH^- \rightarrow HCO_3^-$
6.37	$H_2CO_3 = HCO_3^-$	above reaction continues
6.37-8.35	HCO_3^-	above reaction continues
8.35	HCO_3^- ($H_2CO_3 = 0$)	above reaction completed next reaction begins
8.35-10.33	HCO_3^-	$HCO_3^- + OH^- \rightarrow CO_3^{-2}$
10.33	$HCO_3^- = CO_3^{-2}$	second reaction continues
10.33-14	CO_3^{-2}	second reaction continues

is bicarbonate. If we have a natural water sample at neutral pH, then, we can add strong acid to one fraction and subsequently add strong base to another fraction. The volume of acid required to change the pH from 7 to 5, where all life forms cease to exist, is indicated on the abscissa of Figure 33. Likewise, the volume of base required to raise the pH from 7 to 9, the upper life-limiting pH, is also shown on the abscissa.

In the first case, bicarbonate ion is converted to carbonic acid by the addition of strong acid. In the second case, any remaining carbonic acid is being converted to bicarbonate ion and the bicarbonate ion is being converted to carbonate ion. At a pH of 7, the bicarbonate ion can be converted to the associated acid by the addition of strong acid and the prevailing pH does not change drastically. By the addition of strong base to the system at pH $= 7$, the bicarbonate is converted to the dissociated carbonate anion, and the pH does not change drastically. The presence of the bicarbonate ion, therefore, provides a buffering capacity for the natural water system. In these additions described and illustrated in Figure 33, the following chemical reactions occur:

$$HCO_3^- + H^+ \rightarrow H_2CO_3 \quad \text{(from } 7 > pH > 5)$$

and, again,

$$HCO_3^- + OH^- \rightarrow CO_3^{-2} + H_2O \quad \text{(from } 7 < pH < 9)$$

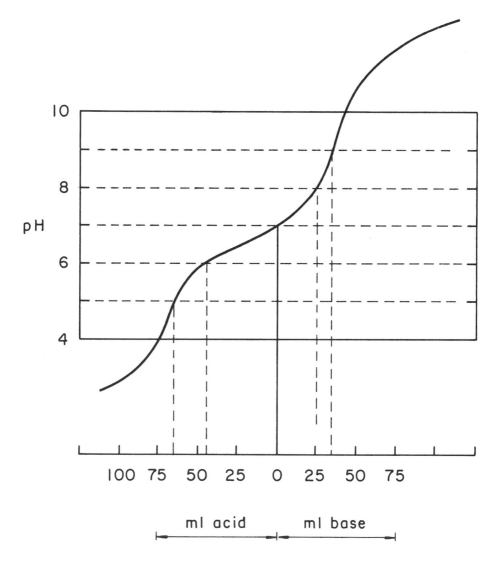

FIGURE 33. Expanded-Scale Carbonic Acid Titration.

With the addition of strong acid, much of the bicarbonate ion must be converted to carbonic acid before the pH drops very much. Upon addition of strong base prior to pH rise above 9, most of the carbonic acid must be converted to bicarbonate ion, and a small fraction of bicarbonate ion, in turn, converted to carbonate ion. With a known carbonate alkalinity concentration, the volume and concentration of an acid or base to be discharged into a natural water can be calculated so that adverse pH effects can be avoided.

EXAMPLE 1. Calculate the equivalents of strong acid which can be added to an industrial settling pond without depressing the pH below 6. The pond has an initial pH $= 7$, carbonate alkalinity $= 10^{-2}$ moles/ℓ, and a 1-day holding capacity.

SOLUTION: Figure 33 presents a detail of the carbonic acid titration curve from Figure 32. The titration curve in Figure 33 within the interval $7 > pH > 6$ indicates that 40.0 ml (0.040 ℓ) of acid cause a 1-unit change in pH. The concentration of strong acid titrant is 0.02 N or 0.02 equivalents/ℓ and the tested water sample is 100 ml (0.100 ℓ).

$$\text{Acid Permitted} = \frac{\left(0.02 \text{ equiv}/\ell\right)(0.040 \text{ } \ell)}{0.100 \text{ } \ell}$$

$$= 0.008 \text{ equiv acid daily}/\ell \text{ pond volume}$$

$$= 8 \text{ equiv acid daily}/m^3 \text{ pond volume}$$

A similar determination from $7 < pH < 8$ indicates that 25 ml (0.025 ℓ) of strong base addition caused a 1-unit change in pH, the comparable calculation follows:

$$\text{Base permitted} = \frac{\left(0.02 \text{ equiv}/\ell\right)(0.025 \text{ } \ell)}{0.100 \text{ } \ell}$$

$$= 0.005 \text{ equiv base daily}/\ell \text{ pond volume}$$

$$= 5 \text{ equiv base daily}/m^3 \text{ pond volume}$$

If such a pond addition did occur, the different carbonate alkalinity forms could be determined by the following procedure:

$$H_2CO_3 \rightarrow HCO_3^- + H^+$$

$$K_{A_1} = \frac{[HCO_3^-][H^+]}{[H_2CO_3]} = 4.3 \times 10^{-7} \text{ moles}/\ell$$

and, rearranging,

$$[HCO_3^-] = \frac{K_{A_1}[H_2CO_3]}{[H^+]} .$$

And, next,

$$HCO_3^- \rightarrow CO_3^{-2} + H^+$$

$$K_{A_2} = \frac{[CO_3^{-2}][H^+]}{[HCO_3^-]} = 4.7 \times 10^{-11} \text{ moles}/\ell$$

and, rearranging,

$$[CO_3^{-2}] = \frac{K_{A_2}[HCO_3^-]}{[H^+]}$$

By summing the carbonate alkalinity forms, and if [Alk] = total carbonate alkalinity concentration,

$$[Alk] = [H_2CO_3] + [HCO_3^-] + [CO_3^{-2}]$$

Therefore, from the above expressions for each carbonate alkalinity form,

$$[Alk] = [H_2CO_3] + \frac{K_{A_1}[H_2CO_3]}{[H^+]} + \frac{K_{A_2}K_{A_1}[H_2CO_3]}{[H^+]^2} \tag{3.8}$$

And, factoring out the $[H_2CO_3]$ term,

$$[Alk] = [H_2CO_3]\left\{ 1 + \frac{K_{A_1}}{[H^+]} + \frac{K_{A_2}K_{A_1}}{[H^+]^2} \right\}$$

Substituting alpha for the term in braces on the right hand side, and then rearranging, yields the following:

$$[Alk] = [H_2CO_3]\, \alpha$$

$$[H_2CO_3] = \frac{[Alk]}{\alpha}$$

and, similarly, from the above expressions for each carbonate alkalinity form,

$$[HCO_3^-] = \frac{K_{A_1}[H_2CO_3]}{[H^+]} = \frac{K_{A_1}[Alk]}{[H^+][\alpha]}$$

and,

$$[CO_3^{-2}] = \frac{K_{A_2}[HCO_3^-]}{[H^+]} = \frac{K_{A_2}K_{A_1}[Alk]}{[H^+]^2[\alpha]}$$

EXAMPLE 2. Determine the concentration of each form of carbonate alkalinity @ pH = 2, 5, 8, 11, and 13, when [Alk] = 10^{-2} moles/ℓ.

SOLUTION: Using the above expressions:

$$\alpha = \left[1 + \frac{K_{A_1}}{[H^+]} + \frac{K_{A_2}K_{A_1}}{[H^+]^2}\right]$$

$$[H_2CO_3] = \frac{[Alk]}{\alpha}$$

$$[HCO_3^-] = \frac{K_{A_1}[H_2CO_3]}{[H^+]} = \frac{K_{A_1}[Alk]}{[H^+][\alpha]}$$

$$[CO_3^{-2}] = \frac{K_{A_2}[HCO_3^-]}{[H^+]} = \frac{K_{A_2}K_{A_1}[Alk]}{[H^+]^2[\alpha]}$$

@ pH = 2, $[H^+] = 10^{-2}$ moles/ℓ and,

$$\alpha = \left[1 + \frac{4.3 \times 10^{-7}}{10^{-2}} + \frac{(4.3 \times 10^{-7})(4.7 \times 10^{-11})}{[10^{-2}]^2}\right] = 1.000043$$

and, therefore,

$$[H_2CO_3] = \frac{10^{-2}}{1.000043} = 0.01 \text{ moles/}\ell,$$

$$[HCO_3^-] = \frac{(4.3 \times 10^{-7}\text{moles/}\ell)(0.01 \text{ moles/}\ell)}{(10^{-2}\text{moles/}\ell)(1.000043)}$$

$$= 4.3 \times 10^{-7} \text{ moles/}\ell, \text{ and}$$

$$[CO_3^{-2}] = \frac{(4.3 \times 10^{-7} \text{ moles/}\ell)(4.7 \times 10^{-11}\text{moles/}\ell)(10^{-2} \text{ moles/}\ell)}{(10^{-2}\text{moles/}\ell)^2(10^{-2})}$$

$$= 2.02 \times 10^{-13} \text{ moles/}\ell.$$

Likewise, after comparable calculations,

@ pH = 5, $[H^+] = 10^{-5}$ moles/ℓ and $\alpha = 1.04$.
Therefore,

$$[H_2CO_3] = 0.01 \text{ moles}/\ell,$$

$$[HCO_3{}^-] = 4.13 \times 10^{-4} \text{ moles}/\ell, \text{ and,}$$

$$[CO_3{}^{-2}] = 1.94 \times 10^{-9} \text{ moles}/\ell.$$

And, @ pH = 8, $[H^+] = 10^{-8}$ moles/ℓ and $\alpha = 44.20$.
Therefore,

$$[H_2CO_3] = 2.26 \times 10^{-4} \text{ moles}/\ell,$$

$$[HCO_3{}^-] = 0.01 \text{ moles}/\ell, \text{ and,}$$

$$[CO_3{}^{-2}] = 4.57 \times 10^{-5} \text{ moles}/\ell.$$

Similarly, @ pH = 11, $[H^+] = 10^{-11}$ moles/ℓ and $\alpha = 245,101$.
Therefore,

$$[H_2CO_3] = 4.08 \times 10^{-8} \text{ moles}/\ell,$$

$$[HCO_3{}^-] = 1.75 \times 10^{-3} \text{ moles}/\ell, \text{ and}$$

$$[CO_3{}^{-2}] = 0.01 \text{ moles}/\ell.$$

Finally, @ pH = 13, $[H^+] = 10^{-13}$ moles/ℓ and $\alpha = 2.03 \times 10^9$.
Therefore,

$$[H_2CO_3] = 4.94 \times 10^{-12} \text{ moles}/\ell,$$

$$[HCO_3{}^-] = 2.12 \times 10^{-5} \text{ moles}/\ell, \text{ and}$$

$$[CO_3{}^{-2}] = 0.01 \text{ moles}/\ell.$$

Concentrations of the three carbonate alkalinity forms are shown in Figure 34. The pH, i.e., $-\log [H^+]$, is shown on the abscissa, and carbonate alkalinity as $-\log [\text{Alk}]$ or p[Alk] is plotted on the ordinate.

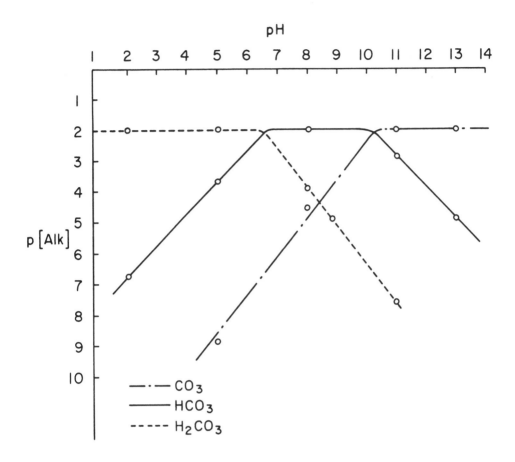

FIGURE 34. p[Alk] - pH Diagram for
Carbonate Alkalinity.

Several interesting concepts are presented here and graphically simplified in this manner.

First, from Example 1, we know that the total carbonate alkalinity
concentration, [Alk] is 0.01 moles/ℓ or 10^{-2}. Therefore, p[Alk] never
rises above 2 on the ordinate, regardless of the pH. Secondly, Figure 34
is a log-log plot, so the concentration lines are linear and the slopes of
these lines can be determined at any pH.

Thirdly, the concentration profile may be read vertically at any
pH. For example, at neutral pH, the carbonate alkalinity profile is as
follows:

$$[H_2CO_3] = 4 \times 10^{-3} \text{ moles/}\ell,$$

82

$$[HCO_3{}^-] = 10^{-2} \text{ moles}/\ell, \text{ and,}$$

$$[CO_3{}^{-2}] = 10^{-6} \text{ moles}/\ell.$$

At pH $= 7$, therefore, $[HCO_3{}^-]$ is only slightly greater than $[H_2CO_3]$. In fact, we know from Table 8 that at pH $= 6.37$, $[H_2CO_3] = [HCO_3{}^-]$. Figure 34 confirms this, because pK_{A_1} is defined by the intersection of the $[H_2CO_3]$ and $[HCO_3{}^-]$ lines on Figure 34. Similarly, pK_{A_2} is shown at pH $= 10.33$. This figure was plotted from the calculations in Example 2. It is very easy to improve the accuracy of such a plot by solving Example 2 with a computer program at very small pH intervals. A homework problem will be assigned to demonstrate this improved accuracy.

4. Hardness

The fate of the hydroxyl ion from the weathering reaction has been discussed in the previous section. What happens to the calcium ion from the weathering reaction? Repeating the reactions,

$$CaCO_3 + 2\,H_2O \rightarrow Ca(OH)_2 + H_2CO_3$$

and

$$Ca(OH)_2 \rightarrow Ca^{+2} + 2OH^-$$

The carbonic acid reacts with the hydroxyl ion to provide a buffering capacity in natural waters, thereby maintaining a near-neutral pH. The calcium ion merely remains in solution and imparts the characteristic of hardness to the water. Natural waters drawn from geographical areas rich in limestone strata, therefore, contain a high concentration of alkalinity and a high degree of hardness. Waters drawn from other strata types are softer waters, but they possess less buffering capacity.

Hardness is measured in units of mg/ℓ $CaCO_3$. Other divalent cations such as Mg^{+2} also contribute to the character of hardness, but to such a smaller extent that they will not be considered here. In order to calculate the hardness concentration, then, the concentration of Ca^{+2} is converted to mg/ℓ $CaCO_3$. This is very convenient, because the molecular weight of $CaCO_3$ is 100 gm/mole, and the equivalent weight is 50 gm/mole. To carry out the conversion of calcium ion concentration to hardness concentration, use the following formula:

$$\text{mg}/\ell \text{ hardness as } CaCO_3 = \frac{(\text{mg}/\ell \; Ca^{+2})(50 \text{ gm/equiv } CaCO_3)}{20 \text{ gm/equiv } Ca^{+2}}$$

83

The calcium ion can be measured by a complexation titration with EDTA, a compound which combines with calcium ion, thereby imparting a color change when excess EDTA is added past the equivalence point. The calcium ion may also be measured by automatic instrumental techniques. The results of these concentration determinations are then converted to hardness units and compared to a hardness scale. Simplistically, a hardness scale is as follows [3]:

soft water contains < 100 mg/ℓ $CaCO_3$ hardness

hard water contains > 100 mg/ℓ $CaCO_3$ hardness

There are varying degrees of hard water and the uses to which the water is put will determine whether the hard water should be softened or not. For example, any industry such as the power-generating industry, which heats large volumes of water, will drive off the dissolved CO_2, raising the pH, converting the bicarbonate alkalinity to carbonate alkalinity, thus precipitating the calcium hardness ion, as indicated in the following reaction:

$$Ca^{+2} + 2\,HCO_3{}^- \rightarrow CaCO_3 + CO_2 + H_2O$$

This precipitated, insoluble $CaCO_3$ will collect on the inside surfaces of pipelines transmitting this heated hard water, inside boilers, heat exchangers, distilling columns, and other equipment in the system until the system becomes inefficient or inoperable. These industries, however, usually remove the hardness before using the water.

In general household cleaning functions, the calcium hardness ion can combine with fatty acids in pure soap formulations and form insoluble precipitates called scum. This is one reason why synthetic detergents have become so popular, because they eliminate these precipitates caused with soap and hardness ion. Some householders still prefer to soften their water regardless of whether they use pure soap or synthetic detergents. The water softening process is usually carried out in a municipal water treatment plant by adding lime, or carried out in the individual household units by installing ion exchange units.

The municipal softening process uses large scale precipitators to remove precipitated $CaCO_3$ produced by the following lime addition reaction:

$$Ca^{+2} \quad + \quad 2\,HCO_3{}^- \quad + \quad Ca(OH)_2 \rightarrow 2\,H_2O \quad + \quad 2\,CaCO_3$$

| hard-ness | alka-linity | lime | precipitated hardness |

84

The above reaction indicates that the accompanying bicarbonate alkalinity, which provides the buffering capacity, is also removed by the chemical softening process. Some state regulations require a certain minimum level of alkalinity be retained in the processed water. Therefore, either the total hardness may not be removed, thereby retaining some hardness and some carbonate alkalinity buffering capacity; or if the total hardness is removed, then the removed alkalinity must be boosted to some minimum level by adding $NaCO_3$. In this case, either the municipal softening process efficiency is reduced, or the cost of replacing the alkalinity may become prohibitive. The tendency, then, even in very hard water areas is to allow the individual householders and individual industries to soften their own water. On an individual household basis, the softening process is normally carried out by ion exchange techniques which will be presented in Chapter 8.

C. Surface Chemistry

Surface chemistry properties, although not well-defined or understood, are very important in environmental engineering. In wastewater treatment, the important surface properties to consider are those on the surfaces of solid particles suspended in water or wastewater. Suspended solids must be measured, as described in Chapter 2, and must be removed to a great extent from the wastewater. Usually these particles are small, difficult to separate from the suspending liquid, and expensive to dispose of. In addition, these suspensions of particles entering a wastewater treatment plant differ in size, density, character of organic and inorganic composition, surface properties, and settleability. For the removal of these suspensions, then, it is helpful to determine at least average values for the above characteristics.

Sizes of these particles transmitted to a wastewater treatment plant are limited only by the diameters of the transmission pipelines. Figure 35 illustrates a particle size spectrum and the related effect of gravity. Large particles usually can be removed by straining. Settleable particles are removed by sedimentation. These two processes are relatively inexpensive. In the region of colloidal sized particles, however, gravity is not an effective removal mechanism, and the suspensions must be treated by rather expensive processes in order remove the particles. Furthermore, 'particles' so small that they are considered to be dissolved are not removed at all in a conventional wastewater treatment plant. Unfortunately, bacteria and viruses fall into these latter categories.

As a wastewater, with its varied sized particles, enters a treatment

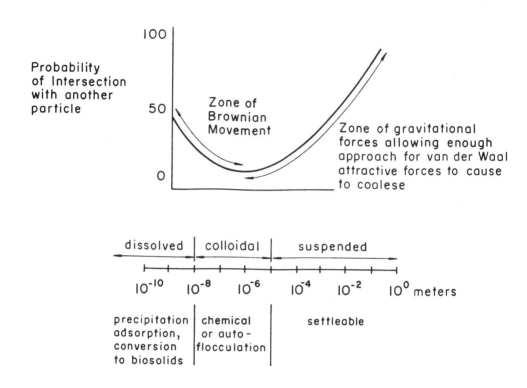

FIGURE 35. Particle Size Spectrum and Gravity Effects.

plant, the easiest particles to remove are taken out of the wastewater stream before damage to downstream equipment is caused. The smaller particles remain, then, and are usually removed as well as possible prior to discharge of the treated liquid. Concentrations of these smaller particles are critical, so it is helpful to look at their surface properties.

Suspensions of large particles are unstable and the particles can be easily removed. When referring to suspensions, it should be noted that 'unstable' is the preferred characteristic! Colloidal and dissolved particles, on the other hand, are stable in suspension and cannot be removed until they are made unstable. What causes this stability of small particles?

Stable particulates are divided into two classes according to their relationship with water. Hydrophobic colloids hate water, and hydrophilic colloids love water as a suspending medium. Hydrophilic colloids are very common in wastewaters since many natural body wastes, soaps and synthetic detergents fall into this classification. They are usually organic materials which have a polar character, like water, with uneven charge distributions. It is almost impossible to remove this class of colloid from wastewater, short of evaporation or some dewatering and drying process. Environmental engineers, in some cases, may need to remove these substances, and many technologies for this purpose are described in Chapter 8. It is simply a matter of cost/benefit, and in almost all cases, the cost of removing hydrophilic colloids far outweighs the benefits.

Hydrophobic colloids, on the other hand, are incompatible with the polar water molecules of the suspending medium. These colloidal particles are so small, however, that they possess a very high surface area to mass ratio. The large relative surface area of inorganic particulates in this size range exhibit a surface charge due to unsatisfied lattice charges in crystalline materials. For organic hydrophobic colloids, the surface charge usually originates from ionized functional groups located on the surface of the 'solid'. The sum of positive and negative surface charges on both organic and inorganic hydrophobic colloids yields a net surface charge. If a colloid has a net negative charge, it will repel other colloids with a net negative charge so that the two particles with like charges can not collide and possibly stick together to gain mass/surface area. A dilute suspension of such colloids results in a stable suspension, since these particles can never gain enough mass for gravity to become effective and cause them to settle out. Figure 36 illustrates such a net negatively charged hydrophobic colloid and also illustrates some further quantitative concepts used in surface chemistry [4]. The negative sites on the surface of this colloid attract positive charges from the hydration layer surrounding the particle in a water suspension. These positive charges can originate from the semi-positively charged hydrogen ion end of a water molecule forming weak hydrogen bonding. The positive charge may also originate from positive cations dissolved in the water, attracted due to opposite electrostatic charges. The association of the net negative surface sites with the positive charges, located in the immediately-surrounding hydration layer, results in a reduction of the net negative surface charge at the outside boundary of this hydration layer.

The outside boundary of this layer is defined by placing the suspension of colloids in an electric field. The negative colloids will migrate toward the positively-charged anode, and the inner hydration layer will migrate with the colloid. The migration rate of the colloids is a function of the surface charge density remaining at the outer boundary of the hydration layer, called the surface of shear.

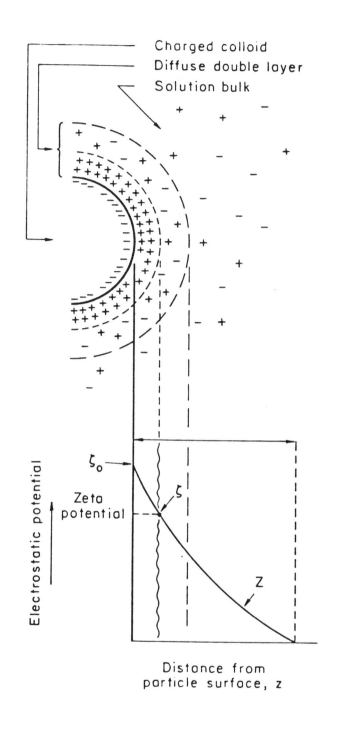

FIGURE 36. Hydrophobic Colloid and Relative Potentials.

88

The migration rate is a function of the applied potential in the electric field, so that the following formula is used to calculate the microelectrophoretic mobility, which is proportional to the zeta potential, the potential at the boundary of the inner, tightly bound layer:

$$\Omega = \frac{dx}{dt}\frac{1}{IR}$$

where, Ω is the microelectrophoretic mobility; x, is the distance traveled by the particle in a measured time period, t; and I and R are the applied current and resistance of the apparatus, respectively.

Except in de-ionized water, the zeta potential measured at the surface of shear, zetz, is always less than the surface potential at the actual solid surface of the colloidal particle zeta$_o$. We can not measure the potential at the actual surface of the colloid using migration techniques because the inner hydration layer, sometimes called the Stern layer, is so tightly bound to the colloid, and migrates with the particle. The zeta potential, therefore, is used as an index of charge density for the hydrophobic colloidal suspension. Figure 36 illustrates how the potential due to the surface charge density diminishes from the high value at the colloidal surface, ζ_o, to the zeta potential at the surface of shear, ζ to lower and lower values, Z, at distances, z, farther away from the colloidal surface. These potentials due to the surface charge density, ζ_o, ζ, and Z are an index of the surface charge density of the colloid and are thus an indication of the tendency of the particles to repel each other and establish a stable suspension. The zeta potential at the surface of shear is the only measurable potential and is therefore the principal measure of the repulsive force on the colloidal particle.

The particles are also subjected to an attractive force, van der Waal's force, which is due to the compatibility of rotating electrons around close atomic nuclei. The closer the atoms of the suspended colloids and their corresponding centroids, the larger is van der Waal's force. This force, then, increases with increasing proximity of the particles, but the magnitude of this force at finite distances away from the colloidal particle surface is much smaller than the negative repulsive force due to like electrostatic surface charges.

The sum of the repulsive force and the attractive force, therefore, results in an energy barrier, shown in Figure 37 [5]. The net negative repulsive force at any distance z from the particle centroids imparts stability to the suspension. Coalescence and any settling due to gravity are consequently prevented. The suspension will remain stable indefinitely unless the net negative surface charge can be overcome.

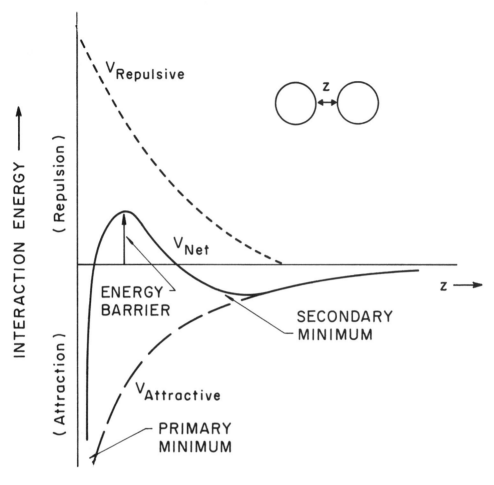

FIGURE 37. Forces on Hydrophobic Colloids.

1. Flocculation and Coagulation

The techniques used to overcome the net negative repulsive force are called flocculation and coagulation. These terms are often used interchangeably but there is a slight distinction between them. Usually, the term coagulation is used to describe the coalescence of particles due to the addition of inorganic chemicals to the stable suspension in order to reduce the net surface charge. Likewise, the term flocculation is usually used to describe the coalescence of particles due to the addition of organic chemicals which bridge between the colloidal particles. These two mechanisms, surface charge reduction and bridging, are used to overcome the net surface charges on the colloidal particles and reduce the stability of the suspension.

2. Surface Charge Reduction

Recalling Figures 36 and 37, we have seen that the repulsive negative force increases with proximity to the colloidal surface. The opposing attractive van der Waal's force likewise increases at very close proximity to the particle surfaces. The resulting net summation of the two opposing forces is largest at the peak energy barrier of the net force, shown in Figure 37. If the particles can be brought close enough together, the energy barrier can be overcome and the particles will enter the portion of the curve with no net negative force, where the particles can coalesce. The principal method for overcoming the energy barrier is to add a concentrated solution of cations, which, with their positive charges, will neutralize the net negative surface charges and allow the particles to approach close enough to overcome the energy barrier. Such a charge reduction effect is shown in Figure 38.

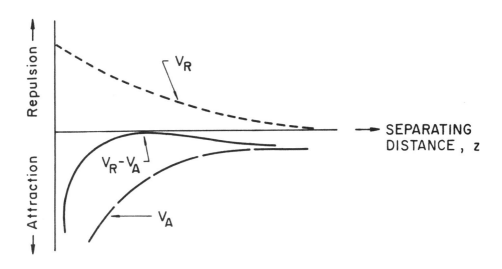

FIGURE 38. Charge Reduction Effect on Surface Forces.

Dosing the stable suspension with cationic coagulant chemicals can de-stabilize the particles by reducing the net negative charge on the surface of the particles. Coagulants with poly-charged cations de-stabilize the suspension more effectively than chemicals with mono-valent cations. The following effectiveness scale is called the Shultz-Hardy series: alum (Al^{+3}) > lime(Ca^{+2}) > soda(Na^+). Furthermore, as can be seen in Figure 37, if too little chemical is added, the suspension remains stable, since the particles can not get close enough to overcome the energy barrier. If too much coagulant chemical is added, the particles are surrounded by excess positive charges, and the net charge on the particles becomes overwhelmingly positive. This dosage overshoot causes an equally stable suspension, this time positively charged, from the originally negatively-charged stable suspension. Any advantage that would have been gained from an increased probability of collision and coalescence is lost and the cost of the chemical addition has been wasted. From this consideration, it should be apparent that there is an optimum dosage for any suspension, and this unique value must be determined empirically for each case.

The necessary chemical dose can best be determined experimentally on a laboratory scale prior to treatment as shown in Figure 39. Here is shown the resulting particle coagulation, increasing with coagulant dosage until the optimum dose is reached and subsequently decreasing after optimum dosage. If the chemical reaction for any precipitation reaction involved in the coagulation process is known, theoretical doses may be approximated from the stoichiometry of the precipitation reaction. But such a case is rare. The requisite concentration of metal-ion coagulant is influenced by such parameters as pH, alkalinity, temperature, mixing rate, and concentration of dissolved and suspended species; for the above salts, the optimum concentration of chemical dose normally is in the range of 150 to 400 mg/ℓ.

EXAMPLE 3. Similar experiments as those illustrated in Figure 39 indicate that 8 ml of coagulant chemical solution containing 20 mg/ℓ is the optimum dosage for a 1 liter suspension sample. Calculate the required dose for a wastewater treatment plant treating 0.05 m³/s.

SOLUTION: By setting up the following proportion:

$$\frac{(0.008 \ \ell)(20 \ \text{mg}/\ell)}{1 \ \ell} = \frac{(\text{unknown dose})}{\left(\frac{0.05 \ \text{m}^3}{\text{sec}}\right)\left(\frac{86{,}400 \ \text{sec}}{\text{day}}\right)\left(\frac{10^3 \ell}{\text{m}^3}\right)}$$

$$\text{Dose} = \left(\frac{691 \ \text{mg}}{\text{day}}\right)\left(\frac{1 \ \text{kg}}{10^6 \ \text{mg}}\right) = 0.7 \ \frac{\text{kg}}{\text{day}}$$

92

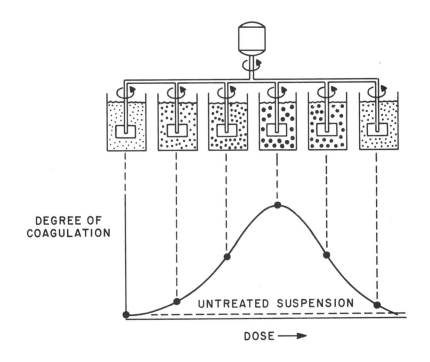

FIGURE 39. Test for Optimum Dosage.

3. Bridging

The bridging mechanism is carried out by long chain polymeric organic molecules called polyelectrolytes. The polymeric chains possess charged functional groups at regular intervals along the chain length. The cationic, non-ionic, or anion charge of the functional group determines the function of each polyelectrolyte. The charge density of the functional groups along the chain length and the pH of the liquid suspension determine the degree of coiling or extension of the polyelectrolyte molecule. For example, if, as before, a negatively-charged hydrophobic colloidal suspension is to be de-stabilized with a cationic polyelectrolyte at a low pH, the resulting excess of hydrogen ions will cause the positively-charged functional groups on the cationic polyelectrolyte to repel each other. (The excess positively-charged hydrogen ion will also have some tendency to neutralize any net negative colloidal surface charges, according to the Shultz-Hardy series.) With excess positive charges, then, the polyelectrolyte chain will be extended, and each end of the polyelectrolyte will more effectively attach to separate, negatively-charged colloidal particles in the stable suspension.

93

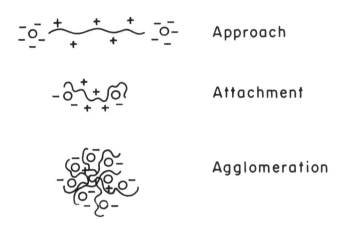

Approach

Attachment

Agglomeration

FIGURE 40. Bridging Mechanism with Polyelectrolyte.

Figure 40 illustrates the progression of the extended polymer chain attaching to two separate particles at each end of the chain and the agglomeration of these separate 'bridged' particles into coalesced flocs. The flocs will have an increased mass/surface area ratio and can settle out due to gravity. In the bridging mechanism case, however, there is no significant surface charge reduction, so the zeta potential remains virtually unchanged, even if the suspension has been de-stabilized.

It should be pointed out that the two cases used to illustrate the charge reduction and bridging mechanisms are very common cases, but only isolated examples in the colloidal suspension spectrum. One must remember that a whole spectrum of cases can be treated from extremely hydrophilic to extremely hydrophobic suspensions. Suspensions can be of inorganic, of organic nature, or of mixed composition. In all of these cases, the suspended particles can be positively- or negatively-charged. The suspensions can be de-stabilized by charge reduction or bridging mechanisms. In addition, if the particles possess a high net surface charge, the pH and the ionic strength of both the suspension and the added chemical have a very strong influence on the treatment efficiency to de-stabilize the suspension. Each suspension must be treated individually, and each optimum dosage determined empirically.

REFERENCES

[1] W. Stumm and J. J. Morgan, *Aquatic Chemistry*, 2nd Edition, Wiley, New York, 1981.

[2] *Handbook of Chemistry and Physics*, CRC Press, Boca Raton FL, 1979.

[3] C. N. Sawyer and P. L. McCarty, *Chemistry for Sanitary Engineers*, 2nd Edition, McGraw Hill, New York, 1968.

[4] *Standard Methods for Examination of Water and Waste-water*, APHS, 14th Edition, APHA, AWWA, and WPCF, 1976.

[5] J. Gregory, *Physical Properties of Sludge*, in *Sludge Characteristics and Behavior*, J. B. Carberry and A. J. Englande, Jr., eds., Martinus Nijhoff Publishers, The Hague, Netherlands, 1983.

HOMEWORK

1. Write the dissociation constant expression for
   a. a weak acid
   b. a weak base

2. If ammonium hydroxide concentration is 0.02 moles/ℓ, calculate the pH of the solution.

3. Calculate the volume of 0.2 N NaOH needed to neutralize 30.0 ml of 0.35 N HCl... and then for 30.0 ml of 0.25 N H_2SO_4. (the normality of H_2SO_4 is half the molarity, since there are two equivalents of H^+ per mole.)

4. Calculate the equivalents of acid permitted and base permitted for discharge in the holding pond described in Example 1 if the pH were allowed to vary between $5 < pH < 9$.

5. An industrial waste with a pH of 3.5 is discharged into the collection system of a municipal wastewater treatment plant at a rate of 0.01 m^3/sec. The average flow of wastewater to the treatment plant is 0.09 m^3/sec, and the municipal wastewater contains 150 mg/ℓ of carbonate alkalinity and a pH of 7.3 Calculate the pH of the combined flow.

6. Calculate the pH of a solution of 0.02 M $NaCO_3$.

7. A buffer solution of 0.1 sodium acetate ($NaC_2H_3O_2$) and

0.1 M acetic acid ($C_2H_4O_2$). Calculate the pH change if 10.0 ml 0f 0.04 M HCl is added to 100 ml of buffer.

8. Calculate the hardness in a water containing 80 mg/ℓ Ca^{+2}. If this water were softened to retain 50 mg/ℓ of bicarbonate alkalinity, how much hardness will remain?

9. Calculate the concentration of remaining Ca^+ in a sludge suspension of $CaCO_3$ if the solubility product of $CaCO_3 = 8.9 \times 10^{-9}$ moles$^2/\ell^2$ when no excess carbonate alkalinity is present. Calculate the Ca^+ concentration in the same sludge suspension when 50 mg/ℓ excess carbonate alkalinity is present

COMPUTER PROBLEM

Calculate the concentration of all alkalinity forms between $2 < pH < 13$ in increments of 0.5 pH units if p[Alk] = 1.5. Plot the results manually or by using a canned plotting program.

Chapter 4

MICROBIOLOGICAL CONSIDERATIONS

A. General Concepts

Almost all waters contain microorganisms. These microorganisms can enter water systems from the atmosphere if they are usually airborne. They can enter underground waters from surrounding soil particles. Drinking water supplies are limited to 1 bacterium per 100 ml as seen in Table 4. Even this water which we consider "safe" to drink can contain bacteria. Bacteria can be eliminated when materials to be made aseptic are heated to 121 °C. at 15 psi pressure. Such procedures are routinely carried out in medical facilities, but they would be much too expensive to conduct at drinking water treatment plants. Public health considerations require that bacteria counts be reduced to reliably low levels at which the probability of disease transmission is minimal. Experience has indicated that such a level of 1 bacterium per 100 ml prevents most disease transmission at a tolerable price. Municipal drinking water plants therefore carry out various processes to remove suspended solids which may or may not be microbes. As a final protection against disease transmission, the water supply is disinfected to this required level. These processes will be treated in detail in Chapter 10.

Once the quality of a drinking water supply is treated to a level which satisfies the Drinking Water Standards, it is distributed under pressure to each municipal household and industry. The water is used in each household or industry and then collected in a separate pipe collection system. The collected wastewater flows by gravity or with intermittent pumping to a wastewater treatment plant. The untreated wastewater contains microbes from human discharges and from stormwater runoff. Most soil microorganisms contributed by surface runoff or underground percolation are harmless or non-pathogenic microorganisms. Pathogenic microorganisms, however, may be contained in discharges from human wastes. If the wastewater is not treated prior to discharge to the environment, the pathogenic microorganisms can survive and perhaps be present when a subsequent drinking water supply is taken from the same source downstream from where the discharge occurred.

Drinking water is treated to comply with Drinking Water Standards, shown in Table 4, and wastewater is treated to comply with discharge permits set for every discharge to the environment. These treatment processes are implemented in order to prevent disease transmission from water borne microorganisms. There are no microbial limits set for wastewater discharges, but the Clean Water Act of 1972 and its subsequent amendments impose ever increasingly stringent limits for suspended solids concentrations present in wastewater discharges. The suspended solids in wastewater treatment plant discharges must be measured as described in Chapter 2 so that an indirect count of bacteria can be obtained by weight.

It should be emphasized that no detection methods or regulation of viruses exist for drinking water supplies or for wastewater discharges. In the United States and Europe, most water borne diseases are very rare and occur only due to unavoidable treatment lapses. In parts of Asia, Africa and South America, water borne diseases are chronic.

B. Qualitative Microbial Identification

Bacterial counts are rarely conducted on either treated or untreated wastewaters, because bacteria are present at such high counts. Instead, the probable removal efficiency is measured by volatile suspended solids measurements, as described in Chapter 2. It is assumed that any bacteria of human origin which escapes treatment and is discharged to the environment will die off quickly in the receiving water. Such die-off rates have been measured as a function of temperature differentials between a normal body temperature of 37 °C. and ambient environmental temperatures [1]. Pathogenic microorganisms which are always present at low concentrations in wastewaters will be reduced to negligible concentrations in the discharge, and those escaping will die off quickly.

It is a common service rendered by health departments, therefore, to detect the presence of human bacteria in rural untreated well-waters, public recreational waters and municipal drinking water distribution systems. Since the existing treatment systems in the United States and Europe are so efficient and since the die-off rate of any surviving bacteria of human origin is so fast, any test results which demonstrate the presence of bacteria of human origin thereby indicates the presence of untreated wastewaters. Usually, in such circumstances, the public health department possesses the power to condemn the well or close the recreational site.

It is important, therefore, that with such power vested in public officials, an acceptable microbial test be adopted which will protect the public health of individuals, but not prohibit the operation of commercial recreational facilities unnecessarily. This is a legally tricky business and involves a risk assessment of available methods, their costs to execute these tests on a routine basis and their probability of detecting pathogenic microorganisms.

Since each and every pathogenic microorganism cannot be tested for and detected at low concentrations, the accepted test identifies the presence of an **indicator microorganism**. The chosen indicator microorganism is the **coliform bacterium** which is present in large numbers in the human intestine and in soils. These bacteria are present in large numbers in untreated wastewaters and the detection of these bacteria in drinking water supplies or in recreational waters indicates the presence of untreated wastewater. Even if the wastewater present in drinking water or recreational water has been diluted by the total volume of the purer water, the detected presence of coliform bacteria increases the probability of the presence of pathogenic microorganisms and their potential for disease transmission.

The adopted coliform test must be carried out under aseptic conditions using sterilized liquid media (broth) and solid media (agar) using sterile glassware and metal wire transfer loops. For each water sample tested, replicate test-tubes must be prepared which contain sterilized lactose or laurel tryptose broth, to contain all necessary growth substrates for coliform bacteria. The broth test-tubes contain an inverted smaller test-tube (called a fermentation test-tube) which will collect any gas such as CO_2 produced by the coliform organisms. A small measured volume of suspected water sample is sterilely introduced into the prepared test-tubes and incubated at 35 °C. for 24 or 48 hours. At the end of the incubation period, the test-tubes are examined for bacterial growth as indicated by the presence of turbidity in the test-tube AND for the presence of gas collected in the inverted fermentation tube. The

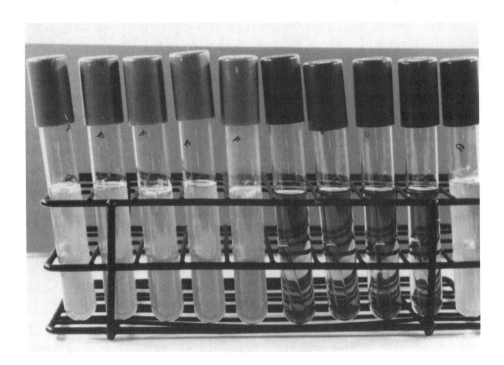

FIGURE 41. Fermentation Results for Coliform Presumptive Test.

absence of turbidity is considered a negative coliform test. The presence of turbidity and the absence of gas in the inverted fermentation tube is considered a negative coliform test. The presence of turbidity and the presence of gas in the inverted fermentation tube indicates a positive Presumptive Test for coliform organisms. An illustration of these variations is shown in Figure 41 [2].

Any tested water sample exhibiting a negative result is now declared uncontaminated and safe. Any water sample exhibiting a positive Presumptive Test must now be subjected to a Confirmed Test. This stage can be conducted in one of two ways: A sterile wire loop can be used to streak an inoculation from a positive Presumptive Test-tube onto an EMB (Eosine Methylene Blue) agar medium in a covered petri dish. Coliform microorganisms develop a particular iridescent colony after incubation for 24 or 48 hours at 35 °C. An illustration of coliform colonies is shown in Figure 42 [2].

The second approved method for conducting a Confirmed Test is to introduce a measured volume of broth from a positive Presumptive Test into a test-tube of Brilliant Green Bile Broth containing an inverted fermentation test-tube. After incubation for 24 or 48 hours at 35 °C., the test-tubes producing growth with gas are considered a pos-

FIGURE 42. Coliform Bacterial Colonies.

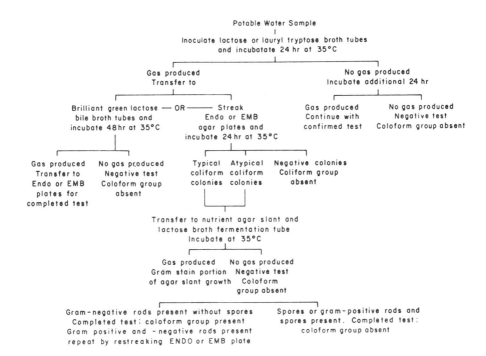

FIGURE 43. Schematic Diagram of Coliform Test.

itive Confirmed Test. Figure 43 illustrates a schematic flow sheet of this series of approved tests for the presence of the indicator coliform organism.

C. Quantitative Enumeration

Once coliform organisms are detected, often it becomes necessary to determine quantitative bacterial counts, of either coliforms present or of total bacterial counts. The type of bacteria counted can be differentiated by choosing a suitable medium. If coliforms are to be counted, EMB agar would be appropriate, since this agar produces visible bacterial colonies of irridescent green which are characteristic of coliforms. If a total bacteria count were desired, a more general agar medium could be used such as nutrient agar.

Three types of enumeration can be carried out. All three very laborious procedures are time-consuming, and must be kept meticulously sterile. Two of the procedures are carried out in a flat petri dish. A top view of a petri dish is shown in Figure 42. It consists of a flat bottom section 150 mm in diameter with a vertical lip about 1 cm in height surrounding its perimeter. The bottom dish is covered by a slightly larger replica serving as a lid, so that the two vertical lips overlap to prevent external contamination. Petri dishes are delivered in sterile condition and must be prepared before enumeration begins by laying a jello-like agar layer in the bottom dish. This semi-solid layer provides the medium with all necessary nutrients upon which the bacteria can grow. For enumeration purposes, one must assume that each single bacterium grows during a one or two day incubation period into one visible colony which can be counted on the medium surface. The laborious aspect involves the water sample manipulation so that several colonies appear on the plate and counting can be carried out accurately.

In order to prepare a water sample for enumeration, a quantitative serial dilution must be made. This dilution consists of taking up 1.0 ml of water to be enumerated into a sterile pipette and introducing this 1.0 ml into 9.0 ml of sterile water in a test-tube. The test-tube contents are mixed vigorously and the sample is now diluted 1:10. If 1.0 ml of the 1/10 dilution is now likewise transferred to 9.0 ml sterile water in a second test-tube, the sample is now diluted 1:100. This procedure can be continued or terminated whenever the examiner estimates that sufficient dilution has been achieved. Figure 44 illustrates this procedure schematically to achieve a dilution to 1:100,000 or 10^{-6}. Once the serial dilution has been carried out to a sufficient degree, then the

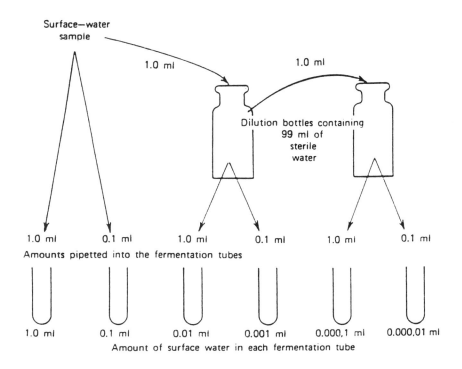

Surface—water
sample

1.0 ml

1.0 ml

Dilution bottles containing
99 ml of
sterile
water

Amounts pipetted into the fermentation tubes

| 1.0 ml | 0.1 ml | 1.0 ml | 0.1 ml | 1.0 ml | 0.1 ml |

| 1.0 ml | 0.1 ml | 0.01 ml | 0.001 ml | 0.000,1 ml | 0.000,01 ml |

Amount of surface water in each fermentation tube

FIGURE 44. Schematic Illustration of Serial Dilution.

plating can begin.

1. PLATE COUNT: A separate sterile pipette is used to withdraw 1.0 ml of diluted sample from an appropriate range of serial dilution test-tubes and each introduced into separate prepared petri dishes. The liquid from each dilution is evaporated at room temperature for a few hours, and then the petri dishes are inverted and incubated in a temperature controlled incubator for 24 or 48 hours. The visible colonies present on the plates are then counted. The plated dilution range should contain both too many visible colonies at a low dilution and too few or zero colonies at the highest dilution. Some intermediate dilution can be chosen, therefore, which produced 20-100 visible colonies to be counted reliably. The visible count is then divided by the appropriate dilution and expressed as counts per milliliter. For example if 45 visible colonies were counted on the plate containing 1.0 ml of the 10^{-4} dilution, the bacterial count will be 45×10^4 bacteria per ml.

2. MEMBRANE FILTER COUNT: This procedure is usually used for water samples which possess low concentrations of bacteria. In addition, this technique is impeded by inorganic turbidity such as from sediments introduced from stormwater runoff. This technique has be-

FIGURE 45. Vacuum Filter Apparatus, Unassembled and Assembled.

come reliable because the process for manufacturing filter papers with specific pore sizes has become routine. A membrane filter composed of polyacetate and containing pores with a diameter of 0.45 μm is used to measure a given volume of water to be tested. For each sample tested, a membrane filter paper is placed on a sterile, perforated filter platform and inserted into a flask fitted with a side arm attached to a vacuum line. A sterile filter cup is clamped onto the membrane and filter platform, the measured volume of water sample is introduced into the filter cup, and the vacuum pressure turned on. The liquid passes through the membrane filter and all particles of diameter greater than 0.45 μm are are retained on top of the membrane filter paper. An illustration of this filter apparatus is shown in Figure 45. Once the filtration is complete, the apparatus is disassembled and the filter membrane paper is removed with sterile forceps, inverted, and placed on an agar-filled petri dish. The paper is inverted so that the captured microbes adhere to the agar to obtain nourishment and grow, each into a visible colony to be counted after incubation for 24 or 48 hours.

Again, the trick is to filter the appropriate volume of sample to be tested so that the visible colonies can be counted. If the water sample to be tested is suspected to be relatively high in microbial concentration, then perhaps 5.0 ml from a serial dilution test-tube can be filtered. The resulting count must be divided by the known dilution factor and the volume of sample filtered in order to convert the number of counted colonies to units of bacteria per ml.

On the other hand, if the water to be tested is known to be relatively clean, large volumes of the undiluted water to be tested can be filtered. For example, liters, tens of liters or hundreds of liters can be filtered in order to obtain a reliable number of colonies to be counted. The results must be expressed in appropriate units, using the smallest volume containing whole bacteria. For example, if 3 liters of water to be tested are filtered and result in 22 visible colonies on the incubated filter paper, the result would need to be expressed as 22 bacteria per 3 liters.

3. MOST PROBABLE NUMBER or MPN : The technique is suitable for testing waters either with low or high bacterial counts, but it results in an indirect statistical prediction of bacterial count, rather than a direct observation and enumeration. For this technique a series of test-tubes is prepared containing sterile broth. Here, a measured volume of 3 serial dilutions must be introduced into 5 replicate test-tubes for each dilution. The test-tubes are capped and then incubated for 24 or 48 hours. The number of test-tubes for each dilution containing bacterial growth observed as turbidity are then counted. If all 5 tubes of one dilution contain growth, the sample is too concentrated. Conversely, if no growth is observed in any of 5 tubes for a given dilution, then the sample has been diluted too much. For intermediate results, the statistical table presented in Appendix 4 must be consulted to provide a statistical result.

EXAMPLE 1. Calculate the number of coliform bacteria if a confirmed test on a potentially contaminated well water resulted in the following turbid indication of growth in dilutions of five replicate tubes of broth.

Sample Size, ml	No. pos.	No. neg
0.1	4	1
0.01	2	3
0.001	1	4
0.0001	0	5

SOLUTION: Referring to Appendix 4, for sample sizes of 10, 1, and 0.1 ml, the series 4, 2, and 1 positive tubes would lead to a result of 26 coliform per 100 ml. Since the dilution of these samples is 100 times greater, then the answer is 2600 coliform/100 ml.

D. Culturing Microorganisms

Consider what happens to microorganisms in a batch reactor, a container which has no flow in or flow out. This type of reactor is used on a small scale operation or in a laboratory test. The reactants are manually introduced into the container, time is allowed for the reaction to occur, and then the final products are manually removed. If such a batch reactor is set up to test the behavior of microorganisms, the following observations can be made: the microorganisms will undergo a period of acclimation, then the population will increase while the microorganisms take up and metabolize the substrate from the liquid phase, and, finally, the population will die off. This sequence is shown in Figure 46. Several time phases in this curve can be identified, and the numbered phases on the curve in Figure 46 correspond to each numbered phase described below.

1. LAG Phase: the microorganisms acclimate to the environmental conditions.

2. EXPONENTIAL GROWTH or LOG GROWTH Phase: the microorganisms grow at a maximum rate, reproducing by binary fission, so that the number, N, equals 2^g, where g is the number of generations. g is defined by the following formula:

$$g = \frac{t_{\text{total}} - t_{\text{lag}}}{t_{\text{generation}}} \tag{4.1}$$

where t_{lag} = acclimation period required before growth begins,
$t_{\text{generation}}$ = time period required for each bacterium
to reproduce by binary fission.

3. ENDOGENOUS RESPIRATION Phase: environmental conditions are no longer conducive to growth, and the rate of growth approximates the rate of death.

4. DEATH Phase: the microorganisms can no longer live in the system due to some adverse condition and the death rate increases.

The following function, based on work by Monod in the 1930's, is used to describe the curve shown in Figure 46.

$$\frac{dX}{dt} = \frac{Y\,dS}{dt} - k_d X \tag{4.2}$$

where S and X = concentration of substrate and

106

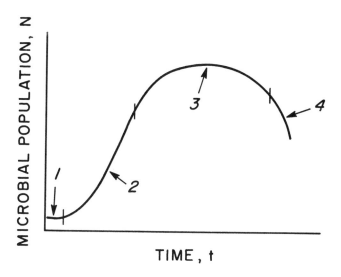

FIGURE 46. Microbial Population Sequence.

microorganisms, respectively, mass/volume;
dX/dt = net change in microbial population,
mass/volume-time;
dS/dt = substrate uptake rate, mass/volume-time;
Y = cell synthesis efficiency or yield, where $Y = -(dX/dt)/(dS/dt)$;
and, k_d = death rate constant, time^{-1}.

In Equation 4.2, the net change in population equals the growth rate (the first term on the right hand side) minus the death rate (the second term on the right hand side). Here, the two-step respiration reaction shown in Figure 21 uses some substrate for both the catabolic reaction and the anabolic reaction. Therefore, an efficiency factor called growth yield, Y, must be included in the Monod model. The factor Y represents the efficiency with which the catabolic respiration in Figure 21 can capture the energy in the ATP high-energy bond.

The sequential growth phases illustrated in Figure 46 can be identified as due to changes in the rate terms of Equation 4.2 as follows:

LAG Phase: $\frac{dX}{dt} \simeq 0, \frac{Y\,dS}{dt} \simeq 0,$ and $-k_dX \simeq 0$

LOG GROWTH Phase: $\frac{dX}{dt} \simeq \frac{Y\,dS}{dt}$ and $-k_dX \simeq 0$

ENDOGEN'S RESP'N: $\frac{dX}{dt} \simeq 0$ and $\frac{Y\,dS}{dt} \simeq -k_dX$

DEATH PHASE: $\frac{dX}{dt} \simeq -k_dX$ and $\frac{Y\,dS}{dt} \simeq 0$

107

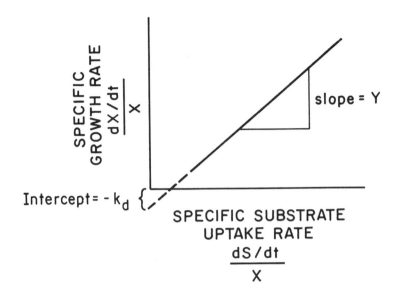

FIGURE 47. Evaluation of Microbial Growth Constants.

If Equation 4.2 is divided by X, Equation 4.3, the **specific net change in population** results:

$$\frac{dX/dt}{X} = \frac{Y\,(dS/dt)}{X} - k_d. \qquad (4.3)$$

This linear function can be plotted, as shown in Figure 47. The characteristic constants Y and k_d can be evaluated numerically or graphically from the plot.

Equation 4.3 can be used to characterize microbial response to any given substrate. Numerical or graphical evaluation of the constants k_d and Y as shown in Figure 47 indicate whether the microbes and substrate tested are metabolically well-suited. Keep in mind that Equation 4.3 was developed for binary mixtures of pure substrates and pure cultures of microorganisms. When engineers adopt these equations for mixed systems of substrates such as wastewaters and impure mixtures of microorganisms such as sludges, only average values for these characteristic constants are obtained.

REFERENCES

[1] J. L. Mancini, "Numerical Estimates of Coliform Mortality Rates Under Various Conditions," *J. Water Poll. Control Fed.*, **43**, 2477, (1978).

[2] M. J. Hammer,, *Water and Wastewater Technology*, John Wiley & Sons, Inc. New York, 1975.

HOMEWORK

1. What is a pathogenic microorganism?

2. What is an indicator organism?

3. Why is an indicator organism used to detect the possible presence of pathogenic organisms?

4. How are viruses detected?

5. Compare the relative effects of drinking water treatment and wastewater treatment on the control of epidemics in developing countries. Should one type of treatment be financed prior to the other?

COMPUTER PROBLEM

Calculate the number of unicellular bacteria grown from a population of 100 at time zero at half hour time intervals until 48 hours. Assume that the lag period is one half hour, the generation time is 20 minutes, and exponential growth takes place for the entire growth period.

Chapter 5

REACTION KINETICS AND REACTORS

A. Kinetics and Mechanisms

Kinetics is the study of reaction rates. Most of you are familiar with cases in which a system is studied at equilibrium. Such examples were examined in Chapter 3 for acid/base dissociation and neutralization. In such thermodynamic studies the following reaction can be used as an example, and this reaction is carried out as long as necessary until reaction is 'complete':

$$aA \ + \ bB \ \rightarrow \ cAB \tag{5.1}$$

When no more reaction occurs, i.e., when equilibrium is reached, the concentrations of A, B, and AB can be measured and used to calculate the equilibrium constant, K_{eq}.

$$K_{eq} \ = \ \frac{[AB]^c}{[A]^a[B]^b} \tag{5.2}$$

The exponents in the equilibrium constant expression correspond to the stoichiometric coefficients in the reaction. But in this case, the time span for this equilibrium to occur is not indicated. Obviously, the experimenter knew how much time elapsed before equilibrium was reached, but this observation was not required for the calculation, so it was not included in the data.

Kinetics, on the other hand, is quite the other way around. Kinetic studies determine the RATE of the reaction, and if the reaction goes to completion so that equilibrium is reached, the study has been carried on longer than necessary to obtain kinetic data. You will see, similarly, when we discuss reactor kinetics (the application of kinetics to the design of reactors) that if the wastewater stream is left in the reactor too long so that the treatment reaction has gone nearly to completion, the engineer has over-designed the reactor.

In order to determine the rate of the reaction in kinetic studies, one could measure one or both of the following rates:

a. the disappearance of the reactant(s)

b. the appearance of the product(s)

In order to design appropriate reactors, the environmental engineer is interested in the former case while the chemical engineer is interested the latter. The following rate expressions indicate the DYNAMICS of the chemical reaction expressed by Equation 5.1:

$$\text{rate forward} \quad = \quad R_f \quad = \tag{5.3}$$

$$\text{a.) } R_{f_A} \quad = \quad - \frac{d[A]}{dt} \quad = \quad K[A]^p[B]^q$$

This expression is read as follows: 'The rate of forward reaction with respect to A equals the decrease in concentration of A with respect to time, which equals the **reaction rate constant** times the concentration of A to the p power times the concentration of B to the q power'. And, similarly for the other reactant,

$$\text{b.) } R_{f_B} \quad = \quad - \frac{d[B]}{dt} = \quad K'[A]^p[B]^q$$

And, lastly, for the product,

$$\text{c.) } R_{f_{AB}} \quad = \quad \frac{d[AB]}{dt} \quad = \quad K''[AB]^z$$

where $[A]$, $[B]$, and $[AB]$ are the measured concentrations of the reactants and product, respectively. The third reaction is positive because AB is produced, not destroyed. The K's are reaction rate constants measured with respect to the indicated component. The exponents indicate the **reaction order** with respect to the indicated reactant or product. For example, we say that the above reaction is p order with respect to A, etc. The exponents in the rate expression do NOT necessarily correspond to the stoichiometric coefficients in the reaction of Equation 5.1. Therefore, you can not look at the stoichiometry of a written reaction and psych out the reaction rate order with respect to any of the reactant(s) or product(s). Instead, the reaction rate constant and order of the reaction must be determined experimentally. For example, we cannot look at the word reaction expressed in Equation 1.3 and conclude that one 'molecule' of substrate combines with one 'molecule' of microorganisms and one 'molecule' of DO to be first order with respect to all the reactants. In fact, we do not know the order or molecularity of Equation 1.3, because it is too complex. It has been expressed, therefore, as a word reaction, not a stoichiometric reaction. We will look only at simple cases, zero order and first order, and how to reduce more complex cases to those simple cases. Even for these simple cases, one cannot look at the stoichiometry of the reaction and assume that any particular kinetic order is appropriate, because the reaction order must be determined experimentally.

B. Reaction Rate Order

The exponents of the reaction rate expression indicate the order of the reaction with respect to that parameter. The above example and its notation can be used to illustrate the cases for classic simple systems:

1.) If $p = 0$, then the reaction rate is not a function of [A], and we say that the reaction rate is zero order with respect to A.

2.) If $p = 1$, then the reaction rate is a function of [A], and we say that the reaction rate is first order with respect to A.

3.) If $p \neq 0, 1$, then the reaction is of higher or more complex order with respect to A, and some more complex reaction mechanism is occurring which is not apparent in the stoichiometry.

An entirely comparable set of circumstances applies to the exponents q and z with respect to reactant B and product AB, respectively.

1. Zero Order

In the zero order system, the reaction proceeds at a rate independent of the concentration of the parameter under consideration. This is often the case when a reactant is present in great excess. If, for example, reactant A were present in great excess while the reaction were monitored with respect to the change in concentration of A, the production of AB could occur at a rate independent of the concentration of A. The following expression would be written:

$$R_{f_A} = -\frac{d[A]}{dt} = K[A]^P \tag{5.4}$$

If $p = 0$, $[A]^P = 1$ and Equation 5.4 reduces to the simple form:

$$R_{f_A} = -\frac{d[A]}{dt} = K. \tag{5.5}$$

Rearranging yields the following integral,

$$\int_{A_0}^{A} d[A] = -K \int_0^t dt$$

and integration yields the following integrand for the zero order case, illustrated graphically in Figure 48:

$$[A] = [A]_0 - Kt \tag{5.6}$$

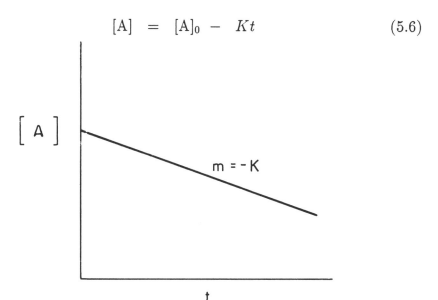

FIGURE 48. Zero Order Kinetic Function.

113

2. First Order

In the first order system, the reaction rate is a function of the concentration of the parameter considered, i.e. the exponent p = 1. The system would follow the rate expression shown in Equation 5.7.

$$R_{f_A} = -\frac{d[A]}{dt} = -K[A] \tag{5.7}$$

Regrouping terms for integration yields the following:

$$\int_{A_0}^{A} \frac{d[A]}{[A]} = -K \int_{0}^{t} dt \tag{5.8}$$

and integration yields the following integrand:

$$\ln \frac{[A]}{[A]_0} = -Kt$$

This equation can be expressed in the following exponential form:

$$\frac{[A]}{[A]_0} = e^{-Kt} \tag{5.9}$$

The minus sign may be eliminated by inverting the ln term of the integrand, as follows:

$$\ln \frac{[A]_0}{[A]} = Kt.$$

And, finally, this function may be expressed in the base 10 system by multiplying K by 0.434, or dividing K by 2.303.

$$\log \frac{[A]_0}{[A]} = kt$$

where $k = 0.434K$. A mixed notation is illustrated in the following expression and in Figure 49.

$$2.3 \log [A] = 2.3 \log [A]_0 - Kt$$

EXAMPLE 1. The following COD data have been collected from a biological reactor with an acclimated culture of microorganisms. The data are a measure of residual biodegradable material in

114

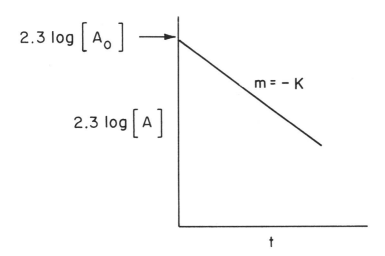

FIGURE 49. First Order Kinetic Function.

the reactor since the COD analysis is used to indicate only changes in concentration.

Time, hr	COD, mg/ℓ
0.0	90.0
1.0	70.0
3.0	50.0
6.0	30.0

Determine the order of the reaction and the reaction rate constant in both the base e and base 10 systems.

SOLUTION: Both linear and semi-log graphs of the data are shown below: Since the linear plot looks like exponential decay, this suggests that these data follow a first order model. The semi-log plot confirms this observation, and the negative slope of the plot is the value of the reaction rate constant, K (base e).

$$\text{slope} = \frac{2.303 \log \text{COD} - 2.303 \log \text{COD}_0}{t}$$

$$= \frac{3.40 - 4.50}{6.0 \text{ hr}} = -\frac{1.10}{6.0 \text{ hr}}$$

$$- \text{slope} = K = 0.18 \text{ hr}^{-1} \quad \text{(base e system)}$$

Since $k = 0.434K$, $k = 0.08 \text{ hr}^{-1}$ (base 10 system)

115

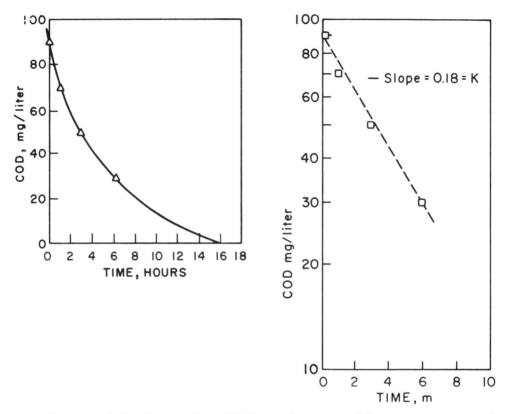

This result indicates that COD analyses, in this case, are a good substitute for the more lengthy BOD test.

3. Complex

If $P \neq 0$ or 1, the system is some complex reaction mechanism which will probably be too difficult for our purposes to analyze mathematically and to apply to the reactor design procedure. Only one complex case will be examined here, the case of enzyme kinetics. Since the wastewater biodegradation reaction is carried out by microorganisms in a biological wastewater treatment plant, and since the microorganisms carry out these reactions with enzyme mediation, it is helpful to examine this case rather thoroughly. Enzyme-mediated reactions are thought to proceed through an intermediate reaction product called a complex intermediate. The complex intermediate can revert to the reactants or it can proceed with another forward reaction to produce the 'final' product and the restored enzyme again. This process is called a consecutive reaction since it is composed of two step-wise reactions. It is illustrated in Equation 5.10 using the conventional notation of S for substrate (or biodegradable material) and E for enzyme.

116

$$S + E \underset{k_{-1}}{\overset{k_{+1}}{\rightleftharpoons}} ES \overset{k_{+2}}{\rightarrow} P + E \tag{5.10}$$

The complex intermediate is symbolized by ES and the product by P. Each arrow can be characterized by reaction rate constants. For example, k_{+1} and k_{+2} characterize the first and second forward reactions, respectively, and those symbols are written above the first and second forward arrows, respectively. Likewise, k_{-1} symbolizes the reaction rate constant characterizing the first reverse reaction, and that symbol is written below the reverse direction arrow. For the reaction to proceed as written, $k_{+2} > k_{+1} > k_{-1}$. We say that the first step is the rate-limiting step. That is, as soon as ES is formed, it quite likely will move right on forward to the products, but it has some tendency to revert back to the reactants. Consequently, the concentration of ES never gets very high, and, in fact, changes very little with time. We can therefore apply the derivation for the case of the "steady state intermediate" to the enzyme reaction above.

The rate may be expressed as a function of any of the reactants or products in Equation 5.10.

$$\frac{d[S]}{dt} = -k_{+1}[S][E] + k_{-1}[ES] \tag{5.11}$$

$$\frac{d[ES]}{dt} = k_{+1}[S][E] - (k_{-1} + k_{+2})[ES] \tag{5.12}$$

and

$$\frac{d[P]}{dt} = k_{+2}[ES] \tag{5.13}$$

If the concentration of ES has achieved a steady state, then

$$\frac{d[ES]}{dt} \simeq 0$$

and Equation 5.12 can be rearranged as follows:

$$[ES] = \frac{k_{+1}[S][E]}{k_{-1} + k_{+2}}. \tag{5.14}$$

Then, since the total concentration of enzyme is present as either free or combined enzyme,

$E_{total} = E + ES$, and thus, $E = E_{total}$ - ES.

This expression for E can be substituted into Equation 5.14, as follows:

$$[ES] = \frac{k_{+1}[S][E_{total} - ES]}{k_{-1} + k_{+2}}$$

if the numerator of this expression is expanded,

$$[ES] = \frac{k_{+1}[S][E_{total}] - k_{+1}[S][ES]}{k_{-1} + k_{+2}} \tag{5.15}$$

If Equation 5.15 then is divided by k_{+1} and the ES terms collected, a modified expression for ES results:

$$[ES] = \frac{[S][E_{total}]}{\dfrac{(k_{-1} + k_{+2})}{k_{+1}} + [S]} . \tag{5.16}$$

This expression can be substituted back into one of the rate equations. Since the rate of reactant disappearance equals the rate of product appearance, Equation 5.11 may be set equal to Equation 5.13. The right hand side of Equation 5.16 can then be substituted into the right hand side of Equation 5.13, as follows:

$$\frac{d[P]}{dt} = -\frac{d[S]}{dt} = \frac{k_{+2}[S][E_{total}]}{\dfrac{(k_{-1} + k_{+2})}{k_{+1}} + [S]} \tag{5.17}$$

Now, k_0, the maximum rate constant, can be substituted for k_{+2}, the largest rate constant described above. Since all of the enzyme is contained in the microbes carrying out the biodegradation reaction, E_{total} can be replaced by X, the concentration of microbes. The complex rate constant in the denominator of Equation 5.17, $(k_{-1} + k_{+2})/k_{+1}$, can be replaced by K_M. This constant is called the half velocity constant and is expressed in units of concentration, usually mg/ℓ. These operations result in the Michaelis-Menten Equation, shown in Equation 5.18 with the previous square brackets removed:

$$-\frac{dS}{dt} = \frac{k_0 XS}{K_M + S} \tag{5.18}$$

If Equation 5.18 is divided by X, the microbial concentration, then the rate of substrate uptake per unit microbe is obtained. This value is known as the **specific substrate uptake rate**, an important characteristic for biological reactors.

118

$$- \frac{dS/dt}{X} = \frac{k_0 S}{K_M + S} \tag{5.19}$$

Several characteristics of this function are shown in Figure 50. The maximum rate constant, k_0, is shown as the asymptotic value of the curve. K_M, the half velocity constant, or often called the Michaelis-Menten constant, is defined as the substrate concentration value when the specific substrate uptake rate is $1/2\ k_0$. It is very difficult to determine these values from the asymptotic curve illustrated in Figure 50. Instead, Equation 5.19 is inverted in order to obtain the linear function given in Equation 5.20 and plotted in Figure 51. This plot is referred to as the Lineweaver-Burke plot and it can be used to linearize asymptotic functions.

$$- \frac{X}{dS/dt} = \frac{K_M}{k_0 S} + \frac{1}{k_0} \tag{5.20}$$

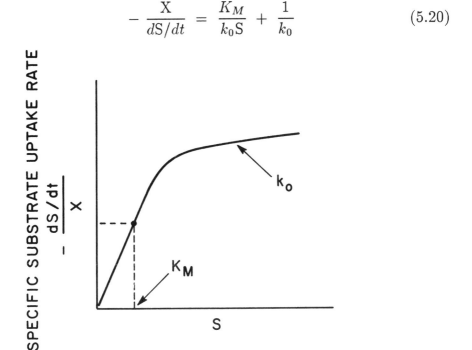

FIGURE 50. Specific Substrate Uptake Rate vs. Substrate Concentration.

If Figure 50 is examined carefully, it is apparent that at large substrate concentrations the specific substrate uptake rate proceeds at a maximum value no matter how much the substrate concentration is increased. Therefore, if $S \gg K_M$, Equation 5.19 can be modified to Equation 5.21, since K_M approaches zero relative to S.

$$\lim_{K_M \to 0} \frac{dS/dt}{X} = -k_0 \qquad (5.21)$$

Now, since Equation 5.19 is modified so that the specific substrate uptake rate is no longer a function of the substrate concentration, we say that the system is 'pseudo' zero order. The asymptotic value in Figure 50 is referred to as the "pseudo-zero order zone" of reaction.

On the other hand, if $K_M \gg S$, Equation 5.19 can be modified to Equation 5.22, since S approaches zero relative to K_M.

$$\lim_{S \to 0} \frac{dS/dt}{X} = -\frac{k_0 S}{K_M} = -kS \qquad (5.22)$$

and the system becomes 'pseudo' first order, where the measured overall rate constant, k, is equal to k_0/K_M. The initial slope section of Figure 50, therefore, is referred to as the "pseudo-first oder zone" of reaction. This latter case is often applicable to biological treatment reactors since the concentration of biodegradable material, S, is low and the microbes are quite acclimated to it. K_M, therefore is relatively high.

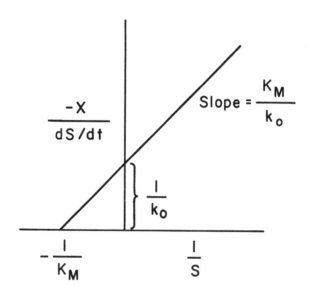

FIGURE 51. Lineweaver-Burke Plot.

EXAMPLE 2. A reactor operated with microorganisms was used to collect the following data. Calculate k_0 and K_M.

$-\frac{dS/dt}{X}, \mathrm{min}^{-1}$	$S, \mathrm{mg}/\ell$
1.35	275.0
1.18	27.5
1.04	14.8
0.69	4.1
0.55	2.8
0.36	1.4

SOLUTION: In order to evaluate the data, a Lineweaver-Burke plot requires an inversion of the data.

$-\frac{X}{dS/dt}, \mathrm{min}$	$\frac{1}{S} (\mathrm{mg}/\ell)^{-1}$
0.74	0.004
0.85	0.04
0.96	0.07
1.45	0.24
1.82	0.36
2.78	0.73

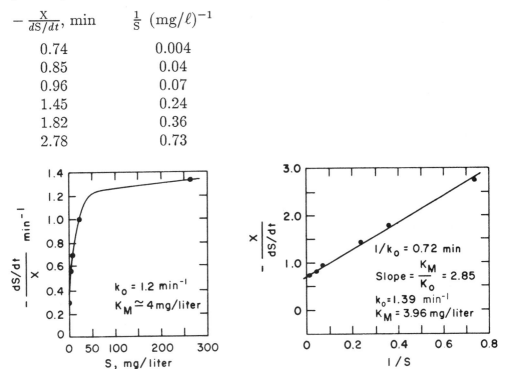

Graphical presentations above indicate the difference between the asymptotic and inverted evaluations. In this case, only a difference in accuracy is apparent between the values obtained by the two different techniques.

Summary of graphical evaluations:

Asymptotic Graph	$k_0 = 1.2 \mathrm{~min}^{-1}$
	$K_M = 4 \mathrm{~mg}/\ell$
Lineweaver-Burke Graph	$k_0 = 1.4 \mathrm{~min}^{-1}$
	$K_M = 4.0 \mathrm{~mg}/\ell$

C. Dependence of Rate Constants on Temperature.

Since chemical and biochemical reaction rates usually increase with temperature, the characteristic rate constant must be temperature-dependent. In the late nineteenth century, the Scandinavian scientist Arrhenius determined that this temperature dependency is an exponential one. Environmental engineers, however, prefer to use the modified power function given in Equation 5.23.

$$k_T = k_{20^\circ C} . \theta^{(T - 20^\circ C.)} \tag{5.23}$$

where k_T is the appropriate rate constant at any temperature, T. This temperature dependency function requires an empirical evaluation of θ, and, for wastewaters in normal temperature ranges, the value $\theta = 1.056$ is often used.

D. Mass Balances

In biological reactors used in wastewater treatment plants, it is somewhat difficult to characterize the system by using any single model. The word reaction in Equation 1.3 indicates that three reactants are present—BOD, microorganisms, and oxygen. Theoretically, it is possible to operate the system so that the BOD, the substrate concentration, S, is the limiting reactant, and a pseudo first order reaction prevails. This modification can be carried out in a large reactor in a conventional biological wastewater treatment plant by loading in high concentrations of microorganisms and dissolved oxygen. Otherwise, the heterogenous mixture of solids (microorganisms), liquid (dissolved BOD) and gas (admixed air as a source of DO) is very complex and is difficult to model mathematically and to actually operate at a treatment plant.

So far, kinetic principles have involved only substrate, or BOD. In order to model the heterogenous mixture in the reactor, the microorganisms must be accounted for as well. In essence, the substrate removed (BOD) will be converted by microorganisms to energy, metabolites (preferably, non-biodegradable metabolites) and new cells via Equation 1.3. The energy and metabolites are not included in the model, but the new cell production is important. All cells in the aeration tank must be settled out in the final sedimentation tank and subsequently recycled back to the aeration tank or wasted and disposed of.

Often, R is substituted for dS/dt or dX/dt, as shown previously in Equations 5.3. The reaction rate, R, is equal to the change in concentration with respect to time. In biological kinetics, it is customary to use subscripts to denote the reactant, e.g., S for substrate, X for the microorganisms. R_S then is the rate of change in the substrate concentration, etc. Equation 5.18 would then read as follows:

$$-R_S = \frac{k_0 XS}{K_M + S}$$

Furthermore, if the rate equation is divided by X in order to obtain the **specific substrate uptake rate**, i.e., the rate per unit of microorganisms, r, is used to denote the specific rate and the same convention is followed with the subscripts. Equation 5.19 for the specific substrate uptake rate becomes

$$-r_S = \frac{k_0 S}{K_M + S} \quad .$$

And, for Equation 4.3, the **specific net change in population**,

$$r_X = Y r_S - k_d \ .$$

These functions have been plotted in Figures 50 and 47, respectively. A similar consideration for the DO can be made; i.e. the specific DO uptake rate can be measured in the batch experiment described above. Equation 5.24 expresses the specific rate function for DO:

$$r_O = \frac{d[DO]/dt}{X} = a\left[\frac{dS/dt}{X}\right] + b \qquad (5.24)$$

where a = DO uptake coefficient for growth
in the anabolic reaction,
and b = uptake coefficient for energy
production and maintenance only
due to the catabolic reaction

These principles can be utilized to model the microorganisms in reactors carrying out wastewater treatment. Now it is possible to write mass balances for appropriate reactants in Equation 1.3. A mass balance expresses the net change in the reactant under consideration, and in these expressions V represents the tank or reactor volume.

MASS BALANCE FOR THE SUBSTRATE:

net rate of change in substrate	=	rate of substrate loaded in	−	rate of substrate discharged	+	substrate uptake rate

$$V\left(\frac{dS}{dt}\right) \quad = \quad QS_{in} \quad - \quad QS_{out} \quad + \quad V R_S \quad (5.25)$$

123

MASS BALANCE FOR THE MICROORGANISMS:

rate of net change in micro-organisms	=	rate of micro-organisms loaded in	−	rate of micro-organisms discharged	+	rate of micro-organisms produced

$$V\left(\frac{dX}{dt}\right) = QX_{in} - QX_{out} + VR_X \qquad (5.26)$$

and, lastly,

MASS BALANCE FOR DO:

rate of net change in DO	=	rate of DO transferred in	−	rate of DO taken up

$$V\left(\frac{d[DO]}{dt}\right) = V\left(\frac{d[DO]}{dt}\right)_{in} - VR_O \qquad (5.27)$$

E. Batch Reactors

In batch reactors, there is no flow in and no flow out, that is $Q = 0$. Therefore, the terms in the mass balance expressions which contain Q become 0 and drop out. For the substrate case, Equation 5.25 can be simplified then to Equation 5.28.

$$V\left[\frac{dS}{dt}\right] = VR_S \qquad (5.28)$$

The result indicates that the net rate of change in substrate concentration equals the rate of reaction, or rate of substrate uptake. Consequently, if reaction is going on, then a steady state will never be achieved. Furthermore, if the reaction has been allowed to go to completion, and no reaction is occurring, the contents have been allowed to remain in the reactor too long for efficient engineering operation.

How can we use this mass balance in a reactor design? The change in substrate concentration can be monitored with time by sequentially removing samples from the reactor and carrying out the BOD analysis on each one. It is highly unlikely in this case that a pseudo-zero order reaction would occur. But we may find that the results would follow a first order reaction, in which case, the right hand side of Equation 5.7 could be substituted for the rate term in Equation 5.28. Here, we can use S for the substrate concentration, instead of [A] as in Equation 5.7.

$$V\left[\frac{dS}{dt}\right] = V(-KS) \tag{5.29}$$

or to the base 10 system, which is usually used in biological kinetics,

$$V\left[\frac{dS}{dt}\right] = V(-kS)$$

where $k = K/2.303$.

Integration of Equation 5.29 yields $\ln S/S_0 = -Kt$. Converting to the base 10 system and inverting to eliminate the minus sign yields the following:

$$\log\frac{S_0}{S} = kt \tag{5.30}$$

which can be re-arranged to provide the time required for a desired degree of treatment:

$$t = \frac{\log(S_0/S)}{k} \tag{5.31}$$

Now if the volume of wastewater to be treated is known, an appropriate reactor size can be provided and the necessary time allowed for the treatment to take place.

EXAMPLE 3. For a first order model, determine the reaction time required to reduce by 90% an initial BOD of 250 mg/ℓ in a batch reactor if the reaction rate constant, $k = 2.5$ day^{-1}.

SOLUTION: Using Equation 5.31,

$$t = \frac{\log(S_0/S)}{k}$$

$$t = \frac{\log\left(\frac{250\text{mg}/\ell}{25\text{mg}/\ell}\right)}{2.5 \text{ day}^{-1}}$$

$$t = \frac{\log 10}{2.5 \text{ day}^{-1}}$$

$$= 0.4 \text{ day} = 9.6 \text{ hr}$$

If the volume of the reactor equalled the volume of wastewater to be treated in a 9.6 hr time period, 90% efficiency would be achieved.

In some cases, kinetic laboratory studies may require the Michaelis-Menten model to be used. In that case, with the neagative sign transposed to the right hand side of Equation 5.18, this term on the right hand side could be substituted into Equation 5.28 for the purpose of reactor design.

$$V\left[\frac{dS}{dt}\right] = -\frac{Vk_0XS}{K_M + S} \tag{5.32}$$

Normally, Equation 5.32 is not used to model a batch reactor system because its integration is rather complex. Instead, the simplifying assumptions shown in Equations 5.21 and 5.22 are implemented so that the system may be simplified to a pseudo-zero order or pseudo-first order reactor.

In addition, the substrate removed is converted to energy and new cells. In order to model the entire system, simultaneous integration of the differential equations for substrate removal and microbial growth should be carried out. The easiest approach to this technique is to use numerical integration computer routines. The Runge-Kutta numerical method for simultaneous integration reprinted in Appendix 5 has been adapted from an example cited in Appendix 5.

F. Continuous Flow Reactors

Continuous flow reactors can be modeled as **completely mixed** or **plug flow**. The completely mixed reactor is characterized by 'instant and complete' mixing of the contents of the reactor. The concentration of BOD loaded into a reactor, then, is 'immediately' diluted by the total volume of the reactor. Of course, the contents are not 'immediately' mixed, but this model of a completely mixed reactor is a theoretical ideal. This model is square-shaped or round in order to facilitate mixing. The completely mixed reactor is an easier mathematical model to use since the concentrations of reactor contents are uniform, the concentration of a component in the reactor is the concentration discharged from the reactor, and steady state can be achieved if the net rate of change is zero.

Plug flow reactors, on the other hand, are characterized by no mixing at all. Conceptually, each differential volume travels intact through the length of the long, narrow reactor. Again, this ideal condition is never achieved, i.e., there is always some mixing of the heterogenous wastewater treatment system. But we say that this mixing causes the system only to deviate from the ideal model. In the plug flow reactor, the contents never achieve steady-state within a differential volume, and complex simultaneous integrations must be carried out as in the batch reactor case. For this reason, only the completely mixed reactor will be used for our modeling purposes.

First, consider the completely mixed reactor without recycle. For the substrate mass balance, Equation 5.25 is used. If a first order rate expression is substituted from Equation 5.7 for R_S, Equation 5.25 becomes

$$V\left[\frac{dS}{dt}\right] = QS_{in} - QS_{out} - VkS$$

Since $Q = V/t$, both sides of the equation can be divided by Q, and then Θ, the theoretical hydraulic detention time, can be substituted for V/Q, the chronological time required to carry out the desired change in S concentration. Since $S_{out} = S$ present in the completely mixed reactor, these two terms are the same and we can eliminate the subscript 'out'. The following expression results:

$$S_{in} - S - \Theta kS = \Theta \frac{dS}{dt}$$

Re-arranging yields a differential equation

$$\int_{S_{in}}^{S} \frac{dS}{S_{in} - S(1 + \Theta k)} = \int_{0}^{t} \frac{dt}{\Theta}$$

and integration, using a technique for singular functions, yields the following integrand:

$$t = \frac{\Theta}{1 + \Theta k} \ln \left[\frac{S_{in}\Theta k}{S_{in} - S(1 + \Theta k)}\right] \qquad (5.33)$$

and

$$S = \frac{S_{in}}{1 + \Theta k}\left[1 - \Theta k e^{-\left(\frac{t}{\Theta}\right)(1 + \Theta k)}\right] \qquad (5.34)$$

As t becomes larger, the exponential term in Equation 5.34 becomes very small, and steady-state is achieved so that Equation 5.34 reduces to Equation 5.35.

$$S = \frac{S_{in}}{1 + \Theta k} \qquad (5.35)$$

If the biodegradation rate constant, k is known, Equation 5.33 can be used to determine an appropriate time to achieve a steady-state condition for a variation in theoretical detention time, Θ. Equation 5.34 can be used to calculate the substrate concentration at any time, t, and Equation 5.35 will indicate the degree of substrate reduction achieved under steady-state operation. Finally, with a known $Q = V/\Theta$, the reactor volume can be calculated.

EXAMPLE 4. Calculate the substrate concentration after 4 hours in a completely mixed reactor without recycle if $\Theta = 6$ hr, $k = 0.6$ hr^{-1}, and $S_{in} = 250.0$ mg/ℓ.

SOLUTION: Using Equation 5.34

$$S = \frac{S_{in}}{1 + \Theta k} \left[1 - \Theta k e^{-\left(\frac{t}{\Theta}\right)(1 + \Theta k)} \right]$$

$$S = \frac{250 \text{ mg}/\ell}{1 + (6 \text{ hr})(0.6 \text{ hr}^{-1})}$$

$$\left[1 - \left\{ (6 \text{ hr})(0.6 \text{ hr}^{-1}) e^{-\left(\frac{4 \text{ hr}}{6 \text{ hr}}\right)\left(1 + (6 \text{ hr})(0.6 \text{ hr}^{-1})\right)} \right\} \right]$$

$$S = \left(54.3 \ \frac{\text{mg}}{\ell} \right)(0.83)$$

$$S = 45 \text{ mg}/\ell$$

If the Michaelis-Menten model is used to design a reactor, again, transpose the negative sign to the right hand side of of Equation 5.18 and substitute this term on the right hand side for the rate term in the substrate mass balance expression. For a complete description of the liquid and solid phases in the reactor, either the first order substrate removal model or the Michaelis-Menten substrate removal model can be used in conjunction with the Monod model for cell population. The Monod model for cell population is given in Equation 4.2. Again Equations 5.25 and 5.26 are used for the mass balances of substrate and biomass, respectively.

$$V\left[\frac{dS}{dt} \right] = QS_{in} - QS_{out} + VR_S \qquad (5.25)$$

$$V\left[\frac{dX}{dt} \right] = QX_{in} - QX_{out} + VR_X \qquad (5.26)$$

128

The Michaelis-Menten rate expression for substrate is Equation 5.18,

$$\mathrm{R_S} = -\frac{k_0 \mathrm{XS}}{K_M + \mathrm{S}} \tag{5.18}$$

and the Monod rate function for the biomass, Equation 4.2,

$$\mathrm{R_X} = Y\frac{d\mathrm{S}}{dt} - k_d\mathrm{X}. \tag{4.2}$$

Substituting the right hand side of Equation 5.18 into the right hand side of Equation 4.2 for $d\mathrm{S}/dt$ yields,

$$\mathrm{R_X} = \frac{Y(k_0 \mathrm{XS})}{K_M + \mathrm{S}} - k_d\mathrm{X}. \tag{5.36}$$

Now to simplify Equation 5.36, a new term is introduced, the maximum growth rate constant,

$$\mu_{\max} = k_0 Y$$

and substituting this term into Equation 5.36 yields,

$$\mathrm{R_X} = \frac{\mu_{\max}\mathrm{SX}}{K_M + \mathrm{S}} - k_d\mathrm{X} \tag{5.37}$$

Now, substituting Equation 5.18 into Equation 5.25 for the substrate mass balance and substituting Equation 5.37 into Equation 5.26 for the biomass mass balance yields the following expressions:

$$V\left[\frac{d\mathrm{S}}{dt}\right] = Q\mathrm{S_{in}} - Q\mathrm{S_{out}} - \frac{k_0 \mathrm{XSV}}{K_M + \mathrm{S}} \tag{5.38}$$

and

$$V\left[\frac{d\mathrm{X}}{dt}\right] = Q\mathrm{X_{in}} - Q\mathrm{X_{out}} + \frac{\mu_{\max}\mathrm{XSV}}{K_M + \mathrm{S}} - k_d\mathrm{XV} \tag{5.39}$$

These equations may be numerically integrated simultaneously using the Runge-Kutta method [1]. Assumptions may be made, however, to simplify the problem. For example, instead of variable operation, steady-state operation can be assumed. The left hand sides of Equations 5.38 and 5.39 become zero in that case. Furthermore, we could compare the concentration of microorganisms in the influent to the concentration of microorganisms in the reactor and decide that the former value is negligible compared to the latter value. The first term on the right hand side of Equation 5.39 becomes zero in that case. Again, dropping the subscript 'out', Equation 5.39 has been simplified to the the following expression at steady-state:

$$0 = -QX + \frac{\mu_{max}XSV}{K_M + S} - k_d XV \qquad (5.40)$$

Then if Equation 5.40 is divided by Q and Θ is substituted for V/Q, the mass balance for biomass yields

$$0 = -X + \frac{\mu_{max}XS\Theta}{K_M + S} - k_d X\Theta \qquad (5.41)$$

Equation 5.41 can be re-arranged to yield an expression for the theoretical hydraulic detention time:

$$\Theta = \frac{1}{\left[\frac{\mu_{max}S}{K_M + S}\right] - k_d} \qquad (5.42)$$

At some minimum critical detention time, the microbes will not have time to reproduce if their generation time has not been exceeded. At detention times less than the minimum theoretical detention time, Θ_{MIN}, the microbes will wash out, no biodegradation will take place, and the discharge concentration, S, will equal the incoming substrate concentration, S_{in}.

$$\Theta_{MIN} = \frac{1}{\left[\frac{\mu_{max}S_{in}}{K_M + S_{in}}\right] - k_d} \qquad (5.43)$$

This value can be used to determine the reactor minimum detention time.

EXAMPLE 5. Calculate the mimimum detention time, Θ_{MIN}, for a completely mixed reactor without recycle used to treat a wastewater with 250 mg/ℓ incoming BOD.

$\mu_{max} = 0.5$ hr^{-1}
$K_M = 50$ mg/ℓ
$k_d = 0.01$ hr^{-1}
$Y = 0.4$

SOLUTION: Using Equation 5.43,

$$\Theta_{MIN} = \frac{50 \text{ mg}/\ell + 250 \text{ mg}/\ell}{(0.5 \text{ hr}^{-1})(250 \text{ mg}/\ell) - 0.01 \text{ hr}^{-1}}$$

130

Equation 5.42 can be re-arranged as follows to provide a function for calculating the discharge substrate concentration as Θ changes:

$$S = \frac{(1 + k_d\Theta)K_M}{\mu_{max}\Theta - 1 - k_d\Theta} \tag{5.44}$$

G. Continuous Flow Reactors with Recycle

For the conventional biological treatment reactor, the model most commonly used is the completely mixed reactor, followed by a secondary sedimentation tank equipped with a recycle line to return microorganisms back to the reactor. Such a schematic is illustrated in Figure 52.

For this case, the mass balance equations must be modified in order to introduce the increased volumetric flow rate from the recycle line and to introduce the biomass from the recycle line. Equations 5.25 and 5.26 are therefore modified as follows for the substrate uptake and biomass

$$V\left[\frac{dS}{dt}\right] = QS_{in} + Q_rS - (Q + Q_r)S + VR_S \tag{5.45}$$

and

$$V\left[\frac{dX}{dt}\right] = QX_{in} + Q_rX_r - (Q + Q_r)X + VR_X \tag{5.46}$$

where Q_r and X_r are the volumetric flow and the concentration of microorganisms in the recycle line, respectively. Here, a new variable is defined as the recycle ratio, $\mathcal{R}$, equal to Q_r/Q. When the mass balances are then divided by Q, Θ is substituted for V/Q, a Monod model is substituted for the rate term, and $\mathcal{R}$ for Q_r/Q, Equations 5.45 and 5.46 become

$$\Theta\left[\frac{dS}{dt}\right] = S_{in} - S - \frac{k_0SX\Theta}{K_M + S} \tag{5.47}$$

and

$$\Theta\left[\frac{dX}{dt}\right] = X_{in} + \mathcal{R}X_r - (1 + \mathcal{R})X + \frac{\mu_{max}SX\Theta}{K_M + S} - k_dX\Theta. \tag{5.48}$$

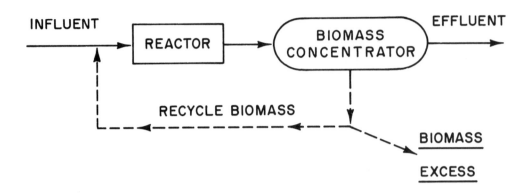

FIGURE 52. Schematic Diagram of Secondary Treatment.

Equations 5.47 and 5.48 can be integrated using numerical Runge-Kutta computer programs to be demonstrated in class. However, if simplifying assumptions are made, steady-state reactor operation may be assumed, the left hand sides of Equations 5.47 and 5.48 become zero, and the first term on the right hand side of Equation 5.48 is assumed to be zero as before. These simplified mass balances for the effluent substrate concentration, S, and the microbe concentration, X, can be solved simultaneously to determine appropriate values of X and Θ to produce a required effluent substrate concentration.

REFERENCES

For a general treatment of reactor kinetics and wastewater treatment, see the following texts. Particular attention should be given to the differing reactor configurations, models, and notation used in each text.

[1] Metcalf & Eddy, *Wastewater Enginering: Treatment, Disposal, and Reuse*, McGraw-Hill Book Company, New York, 1979.

[2] D. W. Sundstrom and H. E. Klei, *Wastewater Treatment*, Prentice-Hall, Inc., Englewood Cliffs, New Jersey, 1979.

[3] E. D. Schroeder, *Water and Wastewater Treatment*, Mc- Graw-Hill Book Company, New York, 1977.

HOMEWORK PROBLEMS

1. Write the reaction rate expressions for the rate constants in the following chemical reaction:

$$aA + bB \rightarrow cC + dD$$

2. Given the following consecutive reaction

$$aA \xrightarrow{k_{+1}} bB \xrightarrow{k_{+2}} cC$$

in which $k_2 > k_1$. Which step proceeds faster? Will the concentration of B increase as the reaction progresses? Which reaction rate constant actually describes the observed reaction rate, k_1 or k_2?

3. Show the algebraic steps used to develop Equation 5.16 from Equation 5.15.

4. The following data were collected from a batch reactor. Determine the reaction rate constant, k.

Time, days	remaining BOD, mg/ℓ
0	180
1	100
2	57
3	32
4	13
5	10

5. Write a mass balance for substrate removal from a completely mixed reactor without recycle and for one with recycle. Explain each term.

6. Use a first order model in a mass balance to calculate the substrate concentration remaining in a batch reactor after 8 hours if the starting BOD concentration $= 280$ mg/ℓ and $k = 2.0$ day^{-1}.

7. Calculate the remaining substrate concentration using a Michaelis-Menten model in a continuous flow, completely mixed reactor at steady state with no recycle if the following constants have been determined for the system:

$$K_M = 50 \text{ mg}/\ell$$
$$k_d = 0.05 \text{ day}^{-1}$$
$$\mu_{max} = 10.0 \text{ day}^{-1}$$
$$\Theta = 6 \text{ hr}$$

8. The following data have been collected from a kinetic experiment. Linearize Equation 5.3a, where the exponent $q = 0$ and determine the value of p, the order of reaction with respect to A.

time Arbitrary Units	[A] Arbitrary Units
10	133
22	68
30	40
40	25
50	14

COMPUTER PROBLEMS

Modify the program in Appendix 5 given for first order substrate biodegradation and Monod microbial growth in a batch reactor to a program for a completely mixed reactor with no recycle. Use the same models for substrate and microbes. Which reactor achieves 90% efficiency of BOD removal faster?

OR

Use Equation 5.33 to calculate the time required to reduce the incoming substrate concentration in Example 4 from 250 mg/ℓ to effluent concentrations, each decreasing by 10% increments. Tabulate results. HINT: Equation 5.33 is a singular function (i.e., it passes through zero at some point).

APPLICATIONS

Chapter 6

PRIMARY WASTEWATER TREATMENT

A. Introduction to Wastewater Treatment

Wastewater treatment plants depend upon adequate collection systems to deliver wastewater to the treatment plant. Frequently, the collection system is inadequate since it was designed to serve less developed areas than those existing now. In addition, often stormwater and domestic wastewaters are transported in 'combined' collection lines which overflow to nearby natural waters if the lines become overloaded during periods of heavy rainfall. In many urban areas, the cost to modify or upgrade collection systems often equals or exceeds the costs to construct new treatment facilities. Wastewater treatment can be, however, only as efficient as the delivery system, and this consideration can not be neglected by planners or engineers.

Modern wastewater treatment plants are designed to utilize biological, chemical, and physical processes in order to remove pollutants. Treatment plants are designed in stages to remove the pollutants in the following order: (1) suspended solids, (2) dissolved biodegradable organic material, and (3) inorganic nutrients. These three stages are referred to as primary, secondary, and tertiary treatment, respectively. Municipal authorities are presently required to implement primary and secondary treatment, and soon may be required to upgrade the degree of treatment to the level of 'best available technology'. Some municipal authorities are presently required to carry out tertiary treatment, if the

136

receiving water basin is particularly fragile.

Compatible industrial wastes may be discharged to the municipal collection and treatment system, and the industry will be required to pay a user charge. Industrial wastes which are not compatible with the municipal system must be treated at the industrial plant site prior to direct discharge to the environment.

The removal of suspended solids must occur initially in the primary treatment stage in order to minimize wear on pumps and subsequent equipment [1]. Removal of soluble biodegradable organic pollutants takes place in the secondary stage, and this stage is the heart of the treatment plant design. This stage minimizes the impact of pollution discharges which would otherwise cause an environmental dissolved oxygen deficit to occur. This removal is accomplished by feeding the polluted water to heterotrophic microbes in an engineered reactor in which the reaction of Equation 1.3 takes place. After the reaction has been carried out, the purified water and microbes are separated and the water sent on for further treatment or discharged to the environment.

Recently, physical-chemical treatment processes have been utilized for removal of these pollutants. Many variations of this type of treatment are being developed today, particularly for industrial wastes. In addition, the processes following secondary treatment often involve physical-chemical processes and no particular order of application exists. Currently, the following levels of pollutant removal must be carried out: (1) greater than 90% removal of suspended solids, (2) greater than 90% removal of BOD, and (3) 80% phosphate removal [2]. A schematic diagram of a typical water treatment plant with no tertiary treatment is shown in Figure 53. Notice that the liquid flows are shown in light lines, while the 'solid' sludge flows are shown in heavy lines.

In this chapter and the following chapters the primary, secondary, and tertiary or advanced treatment stages will be presented in detail. The many options available in each treatment stage will not be combined into a total treatment scheme, as they are in Figure 53. It should be noted that each stage of additional treatment will improve the treatment efficiency, add to the total treatment cost, result in more sludge production, involve more operation and maintenance costs, and probably utilize more energy. These considerations must be included in the management cycle illustrated in Figure 1. In attempting to establish steady-state conditions within a metropolitan area, the engineer or planner must decide how extensive the treatment should be and what costs can be borne by the community in order to maintain the desired level of water quality.

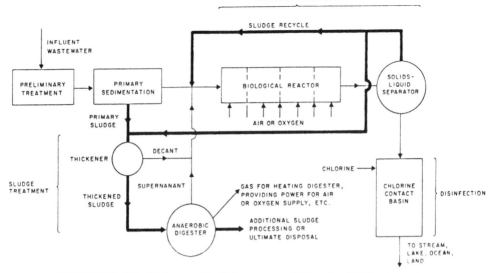

FIGURE 53. Schematic Diagram of Typical Wastewater
Treatment Plant.

B. Preliminary Treatment

Primary treatment consists of a number of unit operations involving
physical processes to remove settleable and suspended solids from the
wastewater. A typical preliminary and primary schematic diagram
is shown in Figure 54. Here it is shown that the wastewater is first
screened to remove large objects which may have entered the wastewater. This preliminary operation is especially necessary in combined
collection systems which contain large diameter pipelines; otherwise,
during storm events, large objects could be washed into the treatment
plant due to increased scouring effect of the large hydraulic flow. The
screens are usually vertical bar screens with spaces between the bars
of 1-3 cm (1/2 - 1-1/2 in). The debris removed in this operation is removed from the bars by scrapers which deposit the debris into hoppers
for disposal.

The next unit operation is the grit chamber which removes abrasive
suspended materials such as sand, coffee grounds, etc. Grit chambers
are designed with a linear flow velocity of 30 cm/sec (1 ft/sec) with a
detention time of 1 minute. These conditions allow the grit to settle
out on the bottom of the chamber where it is collected for disposal.

The wastewater may then flow into a comminuting device which
shreds any material not removed by the previous two operations. Generally, the comminutor consists of a slotted cylinder with rotating cut-
ting members which shear the material to a size small enough to pass

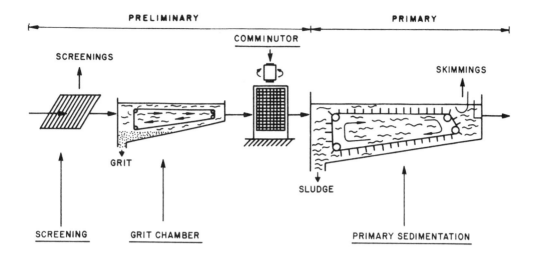

FIGURE 54. Schematic Diagram of Preliminary
and Primary Treatment.

through the slots. Usually the slots are approximately 0.5 cm (1/4 in). In this operation, the sheared solids remain in the wastewater and are subsequently removed in the following operation.

C. Primary Treatment

The primary treatment stage consists chiefly of a primary sedimentation operation. This operation is the most important in primary treatment and is the operation for which the previous preparation has been carried out. Sedimentation tanks are designed to provide relatively quiescent conditions with a detention time of 1-2 hours. The remaining relatively uniform-sized solids will therefore settle to the bottom of the tank where they are pushed by moving scrapers into a sludge hopper for collection and disposal. These conditions permit a removal of 50-75% suspended solids which may account for 40-50% of the biodegradable pollution in the wastewater. An elevated view of a typical rectangular sedimentation tank is shown in Figure 55. The dilute suspension is introduced at the left side, the particles settle to the bottom and are removed from the liquid stream as it travels to the right, and the clarified effluent is removed from the top at the right side.

The suspended particles will settle out under quiescent conditions due to gravitational forces in the downward direction. The downward gravitational force must overcome opposing upward forces due to buoyancy and drag on the suspended particle. The force due to gravity is defined as follows:

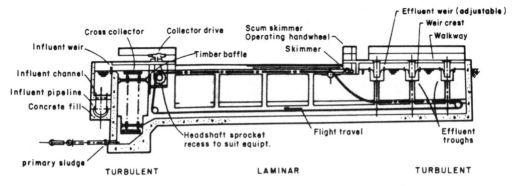

Cross collector   Collector drive   Scum skimmer
Operating handwheel

Influent weir

Timber baffle   Skimmer

Influent channel

Influent pipeline

Concrete fill

Headshaft sprocket   Flight travel   Effluent
recess to suit equipt.   troughs

primary sludge

TURBULENT   LAMINAR   TURBULENT

FIGURE 55. Rectangular Sedimentation Tank.

$$F_G = \rho_p V_p g \qquad (6.1)$$

where V_p = volume of the solid particle
ρ_p = density of the solid particle
and g = acceleration due to gravity

The force due to buoyancy can be defined as follows:

$$F_B = \rho_{liq} V_p g \qquad (6.2)$$

where ρ_{liq} = density of the liquid.
And the force due to drag on the particle can be defined as follows:

$$F_D = \frac{C_D A_p \rho_{liq} v_p^2}{2} \qquad (6.3)$$

where C_D = coefficient of drag
A_p = projected area of the solid particle
and, v_p = settling velocity of the solid particle

These forces can be summed using Newton's Second Law:

$$m\left[\frac{dv_p}{dt}\right] = F_G - F_B - F_D \qquad (6.4)$$

In the above summation, the settling velocity is in a vertical downward direction, and this direction is considered positive. Conversely, the opposing upward forces are considered negative. When the particle achieves its maximum velocity, the particle acceleration, $dv_p/dt = 0$, and Equations 6.1 - 6.3 can be substituted into the right hand side of Equation 6.4, resulting in the following expression:

$$0 = g\left(\rho_p - \rho_{liq}\right)V_p - \frac{C_D A_p \rho_{liq} v_t^2}{2} \tag{6.5}$$

where v_t = terminal, maximum settling velocity, i.e., when $dv_p/dt = 0$.

Rearranging Equation 6.5 to give an expression for the terminal settling velocity, v_t, yields Equation 6.6.

$$v_t^2 = \left[\frac{2g(\rho_p - \rho_{liq})}{C_D \rho_{liq}} \frac{V_p}{A_p}\right] \tag{6.6}$$

Since V_p/A_p is the ratio of volume/area for a solid particle with a diameter d_p, and assuming that the particles are spherical, this ratio may be replaced with $2/3\, d_p$. Furthermore, the Reynold's Number may be defined as follows:

$$N_R = \frac{d_p \rho_{liq} v_t}{\mu} \tag{6.7}$$

where μ = absolute viscosity of the liquid.

Figure 56 illustrates the relationship between the Reynold's Number and the coefficient of drag [3]. It can be seen that at low Reynold's Numbers the dependency of the coefficient of drag is linear. Therefore, the following expression can be used for the coefficient of drag:

$$C_D = \frac{24}{N_R} \tag{6.8}$$

Substituting Equation 6.7 into Equation 6.8 yields the following expression:

$$C_D = \frac{24\mu}{d_p \rho_{liq} v_t}.$$

And substituting this expression and $2/3d_p$ for V_p/A_p into Equation 5.6, yields the following:

$$v_t^2 = \frac{2g(\rho_p - \rho_{liq})}{\rho_{liq}} \frac{2d_p}{3} \frac{(d_p \rho_{liq} v_t)}{24\mu}$$

Simplifying, yields Equation 6.9.

$$v_t = \frac{g}{18\mu}\left(\rho_p - \rho_{liq}\right)d_p^2 \tag{6.9}$$

141

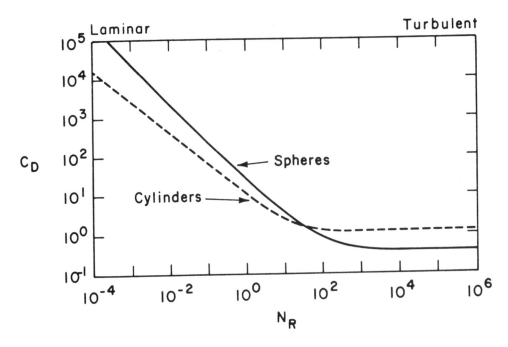

FIGURE 56. Coefficient of Drag as a Function of
the Reynold's Number.

Equation 6.9 was developed by Stoke in the 19th century and is
known as Stoke's Law of Settling. This equation will be used to determine the downward trajectory of a particle under quiescent conditions.
The sedimentation of particles settling separately from dilute suspension is known as Type I settling.

Figure 57 illustrates the streams of laminar flow occurring in a
primary sedimentation tank under quiescent conditions. On the left
hand side of Figure 58 are illustrated the horizontal advective forces
on the solid particles due to the flow through the tank. The horizontal flow, combined with the previously-described vertical downward
motion, shown at the top of the figure, results in a diagonal particle
trajectory in the tank.

A critical particle can be identified which enters at the top of the
tank and settles out on the bottom just before the clarified effluent is
drawn off from the opposite end of the tank. Such a trajectory for a
critical particle is shown in Figure 58.

In designing a sedimentation tank for a required particle removal
efficiency, one must provide quiescent conditions so that the critical
particle will reach the bottom of the tank and be removed from suspension as sludge. Such a critical particle will possess a settling velocity

142

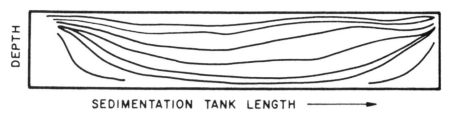

FIGURE 57. Laminar Flow in a Rectangular Sedimentation Tank.

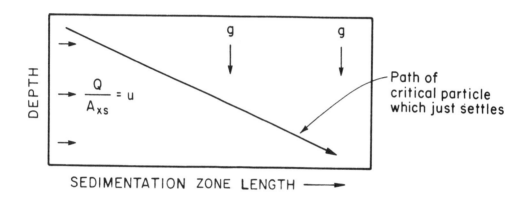

FIGURE 58. Trajectory of a Critical Particle in a
Rectangular Sedimentation Tank.

called the critical settling velocity, v_c. If all particles are uniformly distributed throughout the suspension when it is introduced into the tank, all the particles with a settling velocity equal to or greater than v_c will be removed. Particles with a settling velocity less than v_c will not be removed if they enter the tank at the maximum height. A comparison of these three velocity examples is illustrated in Figure 59.

It is important to realize that the rate at which particles settle out is the same rate at which clarified effluent is produced. In a continuous flow sedimentation tank, then, one can use the critical settling velocity concept in order to design an appropriate tank. If a specific removal is required, say 75% removal of suspended particles, a critical settling velocity can be identified which will provide that removal.

First consider the case of designing a sedimentation tank for a wastewater with a critical settling velocity already determined. In us-

143

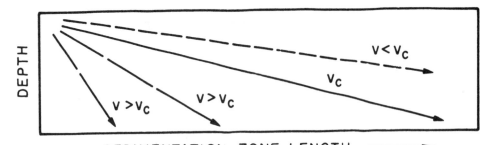

SEDIMENTATION ZONE LENGTH ⟶
FIGURE 59. Particle Settling Velocities in a
Rectangular Sedimentation Tank.

ing this approach, the engineer should keep in mind that the arbitrary
critical settling velocity chosen may or may not be appropriate. The
resulting tank designed from this approach, then, may or may not work
satisfactorily. But for our purposes here, this approach will serve to introduce some important concepts and illustrate the design procedure.

To begin, the engineer should know the value of the volumetric
flow rate, Q, anticipated at the treatment plant. And, if Q is defined
as follows:

$$Q = \frac{V}{t} \qquad (6.10)$$

where V = volume of dilute liquid suspension,
and t = measured time period, or detention time in the tank,

and, since the vertical settling velocity can be expressed as follows:

$$v_t = \frac{\text{tank depth}}{\text{detention time}} = \frac{H}{t} \qquad (6.11)$$

Then the tank detention time can be designed to allow the critical
particle to fall to the bottom of the tank and be removed as sludge. To
do so, v_c is substituted for v_t and the time term from Equation 6.10 is
substituted into Equation 6.11, as follows:

$$v_c = \frac{H}{V/Q} = \frac{H}{(A_sH)/Q} = Q/A_s = q \qquad (6.12)$$

where q = clarification rate, in units of
volume/surface area-time,
and A_s = tank surface area, in unit of length^2.

144

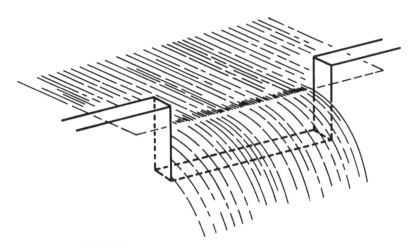

FIGURE 60. Sketch of a Rectangular Weir.

Equation 6.12 states mathematically the concept that the rate of sedimentation equals the rate of clarification. Since the rate of clarification is also the rate at which clarified effluent can be removed from the sedimentation tank, q is often called the overflow rate, because the clarified effluent is removed from the tank by overflowing weirs into collection lines. An illustration of such a weir is shown in Figure 60. Equation 6.12 also illustrates that if the critical settling velocity is known which will provide a specific suspended particle removal and if the anticipated volumetric flow rate for a treatment plant is known, the important design parameter is the tank surface area, A_s.

What is the role of the tank height? The tank height multiplied by the surface area provides the necessary volume for the given volumetric flow rate to remain in the tank for the necessary detention time. If the tank height is very large, then the particles will need to fall a great distance before they reach the bottom and can be removed. If the tank height is very small, the distance that the particles must fall in order to be removed is short, but then the volume provided by the tank may be decreased so much that the detention time in the tank is inadequate to contain the given volumetric flow. In the latter case, the volume of dilute suspension will flow through the tank very quickly and the horizontal advection will be so large that it may scour the settled particles off the bottom into suspension again.

Traditionally, the detention time in a primary sedimentation tank is about 2 hours. Equation 6.10 indicates that this value can be utilized to determine an appropriate depth, as follows:

145

$$tQ = V, \text{ or } (2\,\text{hr})\left(\frac{\text{m}^3}{\text{hr}}\right) = V = A_s H \qquad (6.13)$$

Furthermore, if the surface area of the tank has been determined as previously described, how are the appropriate length and width dimensions determined? Since a sedimentation tank requires quiescent conditions while a continuous horizontal flow is maintained through the tank, tracer studies have shown that a long, narrow tank is most desirable. Such a tank indicates a plug flow tank, characterized by little, if any, mixing. For design purposes, practices have been adopted in which the L/W ratio varies from 3-6:1. Therefore, if A_s has been determined,

$$A_s = (L)(W) = \left[(L:W \text{ ratio})(W)\right] \times (W) \qquad (6.14)$$

then,

$$W = \left[\frac{A_s}{\text{ratio}}\right]^{\frac{1}{2}}$$

and

$$L = \frac{A_s}{W} \quad \text{or} \quad (L:W \text{ ratio}) \times W$$

EXAMPLE 1. For the following wastewater, determine the dimensions of a rectangular primary sedimentation tank with Type I settling:

$Q = 0.044 \text{ m}^3/\text{sec}$ (1 MGD) and $v_c = 3.0$ m/hr (10 ft/hr)

SOLUTION: From Equation 6.12,

$v_c = Q/A_s$, then $A_s = Q/v_c$

$$A_s = \frac{(0.044\,\text{m}^3/\text{sec})(3600\,\text{sec}/\text{hr})}{3\,\text{m/hr}}$$

$$= 52.8 \text{ m}^2$$

If the length to width ratio $= 3$, then $A_s = (3W)(W)$, and

146

$$W = \left(\frac{A_s}{3}\right)^{\frac{1}{2}}$$
$$W = 4.2 \text{ m}$$
$$L = \frac{A_s}{w}$$
$$L = \frac{52.8 \text{ m}}{4.2 \text{ m}}$$
$$L = 12.6 \text{ m}$$

If the detention time $= 2$ hr, or 7200 sec, then

$$V = Qt$$
$$V = (0.044 \text{ m}^3/\text{sec})(7200 \text{ sec})$$
$$V = 317 \text{ m}^3$$

And, since $H = V/A_s$,

$$H = \frac{317 \text{ m}^3}{52.8 \text{ m}^2}$$
$$H = 6 \text{ m}$$

And, furthermore, the overflow rate, q, is calculated as follows:

$$q = \frac{Q}{A_s}$$
$$q = \frac{\left(\frac{0.044 \text{ m}^3}{\text{sec}}\right)\left(\frac{3600 \text{ sec}}{\text{hr}}\right)\left(\frac{24 \text{ hr}}{\text{day}}\right)}{52.8 \text{ m}^2}$$
$$q = \frac{72 \text{ m}^3}{\text{m}^2 - \text{day}}$$

The values calculated above are all within normal design and operational specifications for primary sedimentation.

Since all wastewaters are different in character, all suspended solids in wastewaters may possess differing settling velocities. It is very important to realize that the sedimentation tank design is based on a critical settling velocity value. In order to determine an appropriate critical settling velocity value, a laboratory settling analysis must be carried out on each wastewater to be treated. In Example 1, an appropriate critical settling velocity was provided, but this value may not be appropriate for any other wastewater.

In order to carry out a settling velocity analysis, usually a cylinder with a port is utilized to simulate a quiescent prototype sedimentation tank. The height of the port is measured, and the cylinder filled with a uniform distribution of the suspended solids. At time = 0 a portion of the wastewater sample is removed from the port and a conventional suspended solids analysis is carried out on the sample in order to determine the initial concentration. The suspension is allowed to settle under quiescent conditions, with intermittent samples removed at appropriate time intervals. For each sample withdrawn, another suspended solids analysis must be performed. In this manner, the fraction remaining in suspension at each time interval can be determined. The characteristic settling velocity for each time interval can be measured as H/t', H/t'', H/t''', etc., where the t primes indicate cumulative time intervals. Each corresponding v', v'', v''', must be calculated for these values. The fraction removed from suspension for each time interval will equal (1 - fraction remaining). All particles with $v \geq v'$ will be removed by the end of the time interval t'. All particles with $v \geq v''$ will be removed by the end of the $t''th$ interval, since $t'' > t'$, then $v' > v''$, etc. As t increases, then, particles with smaller and smaller settling velocities will be removed.

Furthermore, some additional particles with a settling velocity less than that characteristic for the particular time interval will be removed if these particles were introduced into the system at a height less than the maximum height. Such trajectories are illustrated in Figure 61. Since 100% of particles will be removed with $v_c = H/t$, those additional particles with $v < v_c$ which enter at a height less than H will be removed in proportion to h/H, where $h < H$. Since h/t equals some $v < v_c$, and $H/t = v_c$, this fractional ratio h/H is proportional to v/v_c. Here in this experimental apparatus we can't measure at which height each individual particle begins its descent, but we can vary t primes to yield the corresponding v primes, all less than v_c. The additional fraction of particles removed, then, which enter the tank at $h < H$ are proportional to v'/v_c, v''/v_c, v'''/v_c, etc. For the first interval, then, the *weight fraction* removed will be $(v'/v_c)X'$, where X' is the *weight* of solids removed in the $'th$ time interval, etc. The sum of the entire incremental fractions removed is given by Equation 6.15.

$$\sum_{i=0}^{i=X_o} = \left[\frac{v'}{v_c}\right] X' + \left[\frac{v''}{v_c}\right] X'' + \left[\frac{v'''}{v_c}\right] X''' \cdots \left[\frac{v^{n'}}{v_c}\right] X_o \qquad (6.15)$$

148

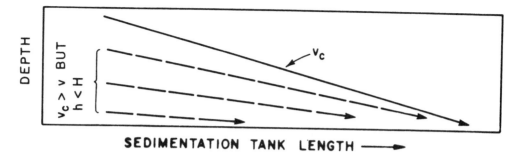

FIGURE 61. Trajectory of Particles with v < v_c
Which Enter at $h < H$.

The total removal, then will be the sum of those particles with a settling velocity equal to or greater than the characteristic settling velocity plus those particles with a settling velocity less than the critical settling velocity if they entered the system at $h < H$. This sum is expressed in words below and mathematically in Equation 6.16

Total Removal	=	fraction of particles with $v \geq v_c$	+	fraction of particles with $v < v_c$ entering at $h < H$

$$\text{Removal}_{\text{total}} = (1 - X_o) + \frac{1}{v_c} \int_0^{X_o} v\,dX \qquad (6.16)$$

where X_o = fraction of particles remaining @ v_c.

The integral may be evaluated graphically or numerically. The numerical method will be demonstrated in class. Example 2 illustrates the evaluation of Equation 6.16 by the graphical method.

EXAMPLE 2. A wastewater experiencing Type I settling is to be analyzed for the design of a primary sedimentation tank to remove 75% of the incoming suspended solids. A laboratory column is set up with a port 1.8 meters below the suspension surface where periodic samples are withdrawn. The initial suspended solids concentration in the wastewater is 200 mg/ℓ. The following data are recorded and used to determine the corresponding values of settling velocity and fraction remaining shown below:

Recorded time	Calculated $v = H/t$	Recorded SS conc.	Calculated Fract'n Remaining
min	m/min	mg/ℓ	X
3	0.61	116	0.58
5	0.37	98	0.49
10	0.18	75	0.38
20	0.09	35	0.18
40	0.05	10	0.05
60	0.03	2	0.01

Equation 6.16 can then be used to obtain a trial and error calculation of the critical settling velocity which will remove 75% of the suspended solids. For a graphical evaluation of Equation 6.16, the fraction remaining data vs. characteristic settling velocity must be plotted as in Figure 62. For simplicity, only the correct trial solution is shown, but the student should attempt other trial solutions at values both less than v_c and greater than v_c. At $v_c = 0.3$ meter/min, Figure 62 indicates that the corresponding y-intercept, X_o, is 0.46, the fraction of particles remaining at that particular settling velocity. This value can be substituted into Equation 6.16, as well as the values for the graphical integration. Since the area under the curve in Figure 62 represents the weight fraction of solids remaining, the area above the curve represents the value for the solids removed. The task then becomes one of integrating the area above the curve from $X_o = 0.46$ to zero. For a trial and error solution such as this, it is easiest to choose dX increments in units of 0.10, while estimating an average v in each dX increment. Multiplying the increment $dX = 0.1 \times v_{average}$ in the increment yields a rectangular area above the curve with equal negative and positive errors between the resulting rectangle and the actual curve. The graphical integration calculations for the trial solution described are as follows:

v	dX	$v \times dX$
.26	.10	.026
.17	.10	.017
.10	.10	.01
.06	.16	.01

$$\sum_{X=0}^{X_o=0.46} v dX = 0.063$$

150

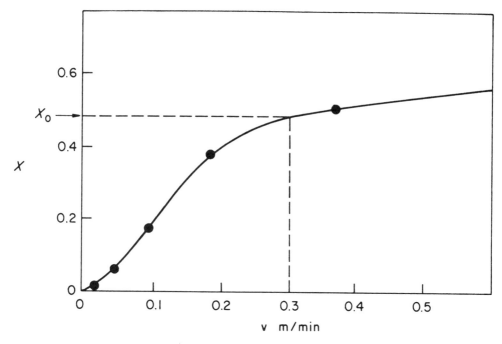

FIGURE 62. Fraction of Particles Remaining vs Settling Velocity.

$$\text{Removal}_{\text{total}} = (1 - X_o) + \frac{1}{v_c} \int\limits_0^{X_o} v\, dX$$

$$\text{Removal}_{\text{total}} = (1 - 0.46) + \frac{1}{0.3}[(0.026) + (0.017) + (0.01) + (0.01)]$$

$$\text{Removal}_{\text{total}} = 0.54 + 0.21$$

$$= 0.75 \text{ or } 75\%$$

The settling velocity chosen then will serve as a critical settling velocity, v_c, which will provide 75% suspended solids removal. This value of 0.3 m/min could then be used to calculate the proper dimensions for the settling tank as shown previously in Example 1.

REFERENCES

[1] G. M. Fair, J. C. Geyer, and D. A. Okun, *Water and Wastewater Engineering*, **Vol 2,** John Wiley and Sons, Inc., New York, 1968.

[2] Metcalf & Eddy *Wastewater Engineering, Treatment, and Disposal,* McGraw-Hill Book Company, New York, 1979.

[3] W. W. Weber *Physicochemical Processes of Water Treatment,* John Wiley and Sons, Inc., New York, 1971.

HOMEWORK PROBLEMS

1. List the types of unit operations performed in the preliminary and primary sections of a wastewater treatment plant and designate the types of particles removed in each case. How do they relate to the particle size spectrum shown in Figure 35?

2. Define Type I settling.

3. What are the vertical and horizontal forces on a particle in a continuous flow rectangular primary sedimentation tank?

4. The design of a primary sedimentation tank is required to have an overflow rate < 65 m^3/m^2 $-$ day by the state. What will be the corresponding critical settling velocity limitation?

5. If the volumetric flow rate $Q = 0.1$ m^3/sec into a new wastewater treatment plant, what would be the required surface area of a primary sedimentation tank using the limits described in Problem 4?

6. If the state required a detention time of 2 hours in the primary sedimentation tank, what tank volume would be required for the treatment plant described in Problems 4 & 5?

7. If design specifications required a length:width ratio of 4:1 in the sedimentation tank designed in Problems 4-6, what would be the dimensions of the tank?

8. What happens to the solids removed in the Preliminary and Primary sections of a wastewater treatment plant?

9. Calculate the total removal expected for Example 2 if v_c = 0.5 m/min and for $v_c = 0.25$ m/min.

COMPUTER PROBLEM

Using SPLINE, SEVAL, and QUANC8 programs provided, calculate the appropriate v_c for 75.0% solids removal of the suspension in Example 2.

OR

Determine the dimensions of a rectangular primary sedimentation tank for Type I settling and calculate the surface overflow rate if $Q = 0.046$ m^3/sec, $v_c = 3.5$ m/hr, and L:W ratio is 4:1. Vary the detention time from 30 minutes to 2 hours in increments of 15 minutes. Tabulate the results with appropriate units.

Chapter 7

SECONDARY WASTEWATER TREATMENT

A. Introduction to Secondary Treatment

The Secondary Treatment stage removes dissolved organic pollutants. This removal may be carried out by biological processes or by less common physical-chemical processes. Biological secondary processes have reached a high level of technology since their development at the beginning of this century, and it is almost always assumed that secondary treatment will be a biological process. These processes imitate previously-described natural processes, and, as such, were thought to be the cheapest type of process to implement. Only recently, costs of certain biological processes have escalated due to costs associated with the following aspects: 1.) biological processes may generate huge volumes of biological sludges which must be treated and transported to a disposal site, and 2.) these highly sophisticated biological processes are usually energy intensive.

Both problems present increasingly prohibitive costs, and the engineer must consider these costs in the overall projection diagrammed in Figure 1. In the future, it may become more cost-effective to implement physical-chemical processes, which are currently more costly to operate than biological systems. This chapter will present the biological treatment processes. And Chapter 9 will describe physical-chemical treatment processes. The latter processes are presently utilized principally as tertiary treatment processes in municipal wastewater treatment plants or for treating specific non-biodegradable wastes in many on-site industrial wastewater treatment plants.

Biological wastewater treatment processes provide appropriate environmental conditions for controlled microbial biodegradation to remove undesired pollutants from a wastewater before it is discharged to the environment. The biodegradable pollutant serves as a food or substrate to the microorganisms. Because of the heterogeneous nature of wastewater, pure cultures of microorganisms never exist in these processes. Rather, a predominate species may develop, or variable species concentrations may be determined by changes in wastewater characteristics and/or environmental conditions in the biological reactor. The metabolic reactions of the various microbial species follow the same reactions as in natural systems shown in Equations 1.3-1.6. In engineered biological reactors, the pollutant is utilized as substrate and is converted to microbially-produced end products or new microbial cells. In order to achieve effective treatment, the microbially-produced end products must be non-polluting or be removed by subsequent treatment stages. Excess new cells produced constitute waste sludge referred to above.

Biological treatment takes place in a controlled reactor designed by engineers to provide satisfactory conditions for the microbes. The reactor may be designed to provide **free suspended** microbes or microbes **attached** to a solid medium. In either case, the reactor is designed to contain the wastewater long enough for the microbes to remove the pollutants from the liquid wastewater and metabolize the pollutants as substrate. When the wastewater has remained in the reactor long enough to become 'purified', the purified liquid and both the old and new microbes must be sent to a final sedimentation tank. Here, the microbes are separated from the 'purified' liquid. This liquid secondary effluent can then be sent to further treatment stages or prepared for discharge to the environment. The microbes, now concentrated into a sludge, may be recycled back to the biological reactor and/or prepared for disposal. It is important to realize that the reactor AND the sedimentation tank together comprise the secondary treatment stage. One tank without the other serves no purpose. Unfortunately, engineers have generally failed to integrate the two steps of secondary treatment, i.e., reaction and separation, and, as a result, secondary treatment often fails. Figure 53 is a typical illustration of secondary treatment.

Biological secondary treatment removes biodegradable carbon prior to wastewater discharge to the environment. Thus, the BOD exerted during the treatment process eliminates the oxygen sag effects described

155

in the Streeter-Phelps Equation. Secondary treatment design is usually based on the volume of the wastewater, known as the hydraulic load, or on the concentration of biodegradable organic material in the wastewater, the BOD load. The major design types are as follows:

1. Lagoons - simply designed and constructed systems, mainly for rural areas with low hydraulic and BOD loads.

2. Trickling Filters - tanks filled with solid rocks or other solid media to which the microbes attach. Wastewater is trickled over the attached microbes. Usually small towns with moderate hydraulic load and low to medium BOD load can use this system.

3. Rotating Biological Contactor - new technology employing large rotating drums of porous microstructure to which the microbes attach. These units are able to reduce sludge production and energy consumption while maintaining high treatment efficiency even with high hydraulic and high BOD loads.

4. Activated Sludge - Tank filled with free suspended microbes thoroughly mixed with the wastewater in order for the biodegradation reaction to occur. The effluent may be further treated or discharged from the plant. This high tech system is used in metropolitan areas with high hydraulic and high BOD loads.

Numerous types of process modifications have been developed for each of the above design types. Generally, however, activated sludge processes are capable of achieving the greatest degree of treatment efficiency, but since this system requires the greatest operational control, it is the most costly to operate. In contrast, lagoons are the least efficient system but are the easiest and least expensive to operate.

B. Design of Secondary Process Reactors

1. Lagoons

Lagoons are usually constructed as fairly shallow, excavated basins. The term 'lagoon' is commonly used interchangeably with 'stabilization pond' or 'oxidation pond'. The process utilizes both heterotrophic and photo-autotrophic microorganisms working together as shown in Figure 63. Heterotrophic microorganisms stabilize the biodegradable organic material in accordance with Equation 1.3. The photo-autotrophs supply oxygen to the heterotrophs in excess of that provided by reaeration from the atmosphere at the lagoon surface and remove inorganic nutrients in accordance with Equation 1.4.

Waste stabilization ponds usually develop facultative systems, as shown in Figure 63, since aerobic conditions exist near the surface and throughout most of the pond depth. Because of the presence of settleable organic material, however, anaerobic conditions may persist near the bottom and metabolic reactions in this zone would follow the reactions in Equations 1.5 and 1.6. Depending on the design, a shallow pond might be, for all practical purposes, completely aerobic; on the other hand, a pond with an overload of BOD might be almost totally anaerobic. Oxidation ponds have been used successfully as single units or in series of multiple units for the treatment of both municipal or industrial wastewaters. Oxidation pond efficiencies are extremely temperature-dependent. In the past, little consideration has been given to sludge separation and disposal at these sites, but recent government regulations have required upgrading these deficiencies to comply with wastewater treatment policies in The Clean Water Act of 1972.

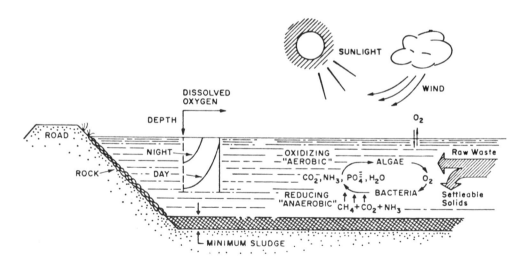

FIGURE 63. Cross-section of a Typical Lagoon.

2. Trickling Filters

Trickling Filters are circular tanks approximately 1 to 4 meters in depth, filled with spherically-shaped rocks or plastic medium. A vertical axle, supporting one or two horizontally-extended shafts fitted with nozzles, is rotated while spraying wastewater on top of the rock bed. Wastewater trickles down onto the top of the bed and then percolates down through the entire depth of the bed. A cutaway drawing of a trickling filter is shown in Figure 64.

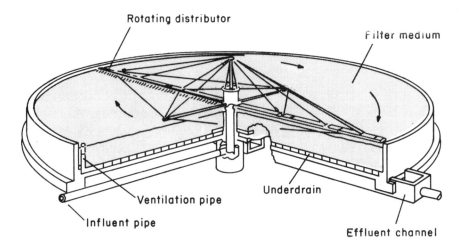

FIGURE 64. Cutaway View of a Trickling Filter

The flowing wastewater provides substrate for growth of aerobic bacteria growing on the rock surface or plastic medium. This growth of heterotrophic bacteria is called bioslime, and it grows in a fixed film covering the solid surface of the filler material. The biodegradable substrate diffuses into the film where aerobic metabolism occurs according to Equation 1.3, and the metabolic products diffuse back out of the film into the liquid stream. This exchange is shown in Figure 65.

The exchange continues with the film growing thicker and thicker until the film can no longer adhere to the solid surface. The film sloughs off in sections due to the hydraulic shear from the flowing wastewater. These large masses of bioslime are then temporarily admixed with the liquid flow until they are separated out in the final sedimentation tank. Trickling filters are classified as high-rate or low-rate filters, depending on the hydraulic load or the BOD load. Figure 66 illustrates various schematics of low and high rate trickling filters.

The low rate trickling filter scheme contains only one reactor, It is a relatively simple device, highly dependable if the hydraulic load

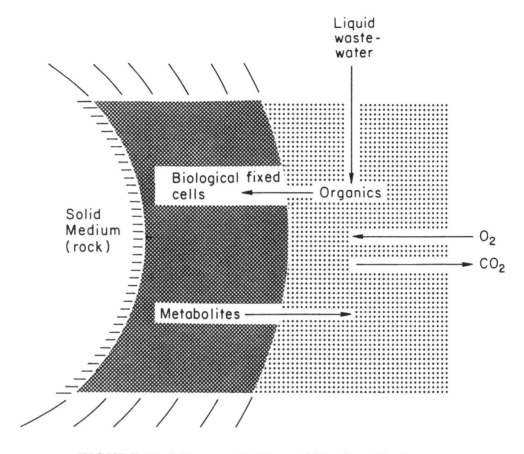

FIGURE 65. Microscopic View of Bioslime Exchange.

is great enough to provide continuous or regulated intermittent waste-water application to the bed. If the hydraulic load is inadequate, and unregulated intermittent application to the bed occurs, the bioslime may dry out and die; the resulting operation would be inefficient and unreliable. In such cases, often the effluent is recycled back through the filter again in order to provide an adequate hydraulic load. Three low rate recycle schemes are illustrated in Figure 66.

Design of trickling filter units are most often based on empirical design parameters correlated from past operational performance. As stated above, the most common empirical design parameters are hydraulic loading and BOD loading. Table 9 provides these values for the various trickling filter categories. Although Table 9 provides design specifications for both low and high rate trickling filters, we will utilize the low rate design procedure only.

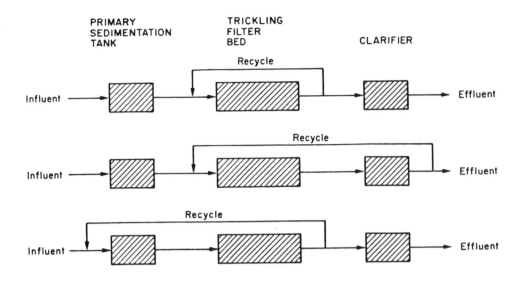

Low Rate Trickling Filter Schemes
One Reactor

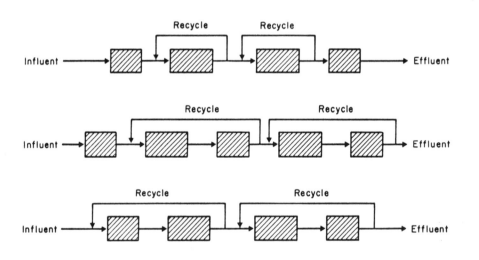

High Rate Trickling Filter Schemes
> 1 Reactor

FIGURE 66. Schematic Diagram of Low and High Rate
Trickling Filters.

Table 9

Trickling Filter Design and Operation Factors

Parameter	Low Rate	High Rate
Hydraulic Load	$1-4$ m^3/m^2-day	$10-40$ m^3/m^2-day
Organic Load(BOD)	$0.08-0.3$kg/m^3-day	$0.3-1.0$ kg/m^3-day
Depth	1.5-3.0 m	1.0-2.0 m
Recirculation	none	1-4
Dosing interval	intermittent	continuous

If an empirical design is to be carried out, it becomes a simple matter to design a unit when the anticipated hydraulic rates and BOD concentrations are known. Normally, the filter volume is calculated on both bases and the larger value is stipulated. See example 1.

EXAMPLE 1. Calculate the volume of a low rate trickling filter required for a wastewater if the volumetric flow rate, Q, is 0.044 m^3/sec and the BOD concentration is 150 mg/ℓ.

SOLUTION:

ON THE BASIS OF HYDRAULIC LOADING:

Most conservative:

$$\text{Area} = \frac{\left(0.044 \text{ m}^3/\text{sec}\right)\left(86,400 \text{ sec}/\text{day}\right)}{1 \text{ m}^3/\text{m}^2 - \text{day}} = 3802 \text{ m}^2$$

Least conservative:

$$\text{Area} = \frac{\left(0.044 \text{ m}^3/\text{sec}\right)\left(86,400 \text{ sec}/\text{day}\right)}{4 \text{ m}^3/\text{m}^2 - \text{day}} = 950 \text{ m}^2$$

ON THE BASIS OF BOD LOADING:

Volumetric Flow Rate $= Q$, and

$$Q = (0.044 \text{ m}^3/\text{sec})(86400 \text{ sec}/\text{day}) = 3802 \text{ m}^3/\text{day}$$

BOD Loading Rate $= Q \times$ BOD

$$\text{BOD Loading Rate} = \left(\tfrac{3802 \text{ m}^3}{\text{day}}\right)\left(\tfrac{150 \text{ mg}}{\ell}\right)\left(\tfrac{1000 \text{ }\ell}{\text{m}^3}\right)\left(\tfrac{\text{kg}}{10^6 \text{ mg}}\right)$$

$$= 570 \text{ kg/day}$$

161

$$\text{Most conservative}: \quad \text{Volume} = \frac{570 \text{ kg/day}}{0.08 \text{ kg/m}^3 - \text{day}}$$

$$= 7128 \text{ m}^3$$

$$\text{Least conservative}: \quad \text{Volume} = \frac{570 \text{ kg/day}}{0.3 \text{ kg/m}^3 - \text{day}}$$

$$= 1901 \text{ m}^3$$

Which design would you choose with such a wide range of results? First, the calculations from the hydraulic loading factor lead to results in units of area, while the calculations on the basis of BOD loading lead to results in units of volume. If we multiply the most conservative result for the area by the most conservative depth, 3 meters, 11000 m³ is obtained. Likewise, if the least conservative area result is multiplied by the least conservative depth value, 1.5 meters, 1434 m³ results. These operations result in two extreme values.

Why not choose some intermediate value permitted for the depth, perhaps 2 meters? Then, we can multiply this depth value times the area values obtained from the hydraulic calculation and, likewise, we can divide the volume values obtained from the BOD loading by this depth value; we can thereby compare the resulting calculations.

	Hydraulic Loading Results		BOD Loading Results	
	Area ($\times$ 2 m)	Volume	Area ($\times$ 2 m)	Volume
Most Conservative	3802 m²	7603 m³	3564 m²	7128 m³
Least Conservative	950 m²	1901 m³	950 m²	1901 m³

These values indicate that the design should be based on the hydraulic loading factor which yields values slightly greater than those from the BOD loading factor. Whether the most or least conservative values should be specified is a matter of policy of the governmental regulatory agency and design firm.

In order to eliminate the large deviation in the most conservative and least conservative values seen above for the design of trickling filter units, quite a few semi-empirical approaches have been proposed. One such technique developed by Eckenfelder [1] uses the following expression shown in Equation 7.1.

$$\frac{S_{effluent}}{S_{influent}} = \exp\left[-\frac{Kh}{q^n}\right]$$ (7.1)

where S values indicate substrate concentration, measured as BOD. The K value is a modified BOD rate constant which is determined experimentally and accounts for the rate of the actual biodegradation reaction, as well as the rate of transport of substrate and oxygen into the slime layer. The h term is the filter bed depth and q is the volumetric application rate per unit surface area, Q/A_s. The exponent n is a constant related to the specific components of the system. If q and h can be varied in a laboratory reactor model, the constants can be evaluated to determine any pre-specified effluent concentration. See Example 2.

EXAMPLE 2. A 3 meter laboratory model trickling filter is used to test a wastewater with BOD = 150 mg/ℓ in order to design a prototype trickling filter. Here in the lab model q varies at hydraulic loading rates from 6.0 to 17.3 m³/m² – day and there is a port for sample withdrawal every 0.75 m in height. During the experimental determination of characteristic K and n values for the particular wastewater, wastewater is continually applied to the laboratory column at the hydraulic loading rates shown below, and when the system has achieved steady state operation, liquid samples are withdrawn simultaneously from each port, and remaining substrate concentration tested by carrying out BOD tests. Equation 7.1 can be used to calculate the proper design alternatives for an appropriate treatment efficiency.

SOLUTION. The following data were obtained from the laboratory model:

Fraction BOD Remaining, S_e/S_i

Depth, meters	$q = 6.0$	12.1	17.3
	m³/m² – day		
0.75	0.6	0.7	0.7
1.50	0.4	0.5	0.6
2.30	0.2	0.3	0.4
3.00	0.1	0.2	0.3

In order to linearize the evaluation of Equation 7.1, plot values of each side of Equation 7.1, ln S_e/S_i vs. h. See Figure 67. The slopes of the plots equal - K/q^n.

163

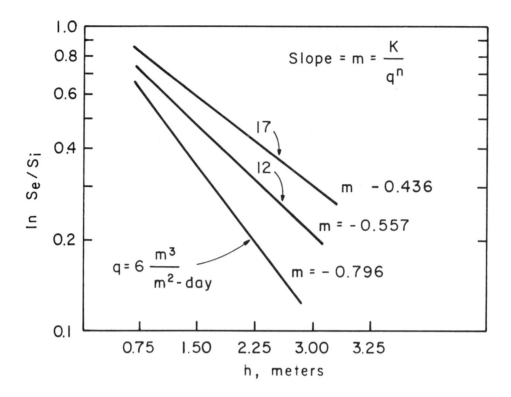

FIGURE 67. Ln S_e/S_i vs Height.

The resulting log of each slope at every q can then be plotted vs. log q as in Figure 68 in order to evaluate K and n, since

$$\log \text{slope} = \log K - n \log q.$$

From Figure 68, K will be the intercept at $q = 1$ (where log $q = 0$) or K = 2.21 day^{-1}. The slope is $-n$, so that $n = 0.57$.

Substituting these values into Equation 7.1 will lead to a trial and error solution for the best design. A programable calculator or a computer makes this an easy exercise. As an example, the previous least conservative value of hydraulic load for a low rate trickling filter from Table 9 can be tested as follows:

$$\ln S_e = (\ln S_i) - \left[\frac{Kh}{q^n}\right]$$

and from Table 9, $q = 4$ m^3/m^2 − day   and   h = 2 meter,

164

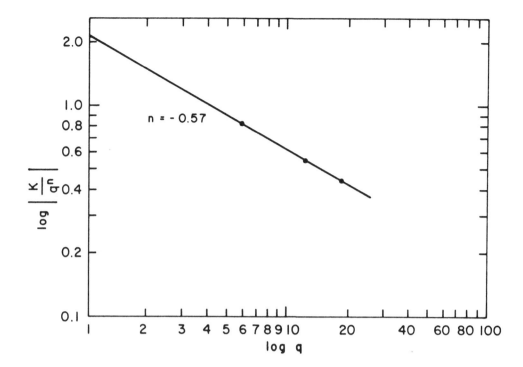

FIGURE 68. Linearization for K and n Determination.

$$\ln S_e = (\ln 150) - \frac{(2.21 \text{ day}^{-1})(2 \text{ m})}{(4 \text{ m}^3/\text{m}^2 - \text{day})^{0.57}}$$

$$= (5.01) - (2.00)$$

$$S_e = 20 \text{ mg}/\ell$$

The result calculated from this data, would indicate that the previous calculation in Example 1, labeled 'least conservative' would be adequate, if the treatment plant were required to carry out 85% efficiency. We could proceed with the design at 950 m² for the surface area and a depth of 2 meters. If the treatment plant were required to operate at a higher efficiency, the calculation could be modified by varying q or h in order to obtain the desired efficiency. A computer problem at the end of this chapter illustrates this determination.

If the design were inadequate at all values of the trial and error solution, the design could be modified to a high rate trickling filter system. In this scheme, trickling filter beds in series are used with some variation in recycle, as shown in Figure 66.

165

3. Rotating Biological Contactors

Rotating Biological Contactors (RBC's) are relatively new, innovative fixed film reactors. The same principles are used as for trickling filters, except the operation is different. In trickling filters, the medium is stationary and the wastewater moving. In the RBC case, the medium rotates in relatively stationary wastewater. These units are in the shape of a rotating horizontal drum constructed of units of porous plastic media in various geometries. The drum is supported above a cylindrical trough up to a depth near its axle. As the drum rotates, each surface area unit is alternately dipped into the wastewater trough and then exposed to the air. The film fixed to the rotating surface, then, alternately receives substrate and an oxygen supply. The biodegradation reaction occurs in a similar fashion to the Trickling Filter. The Rotating Biological Contactor is usually referred to as an RBC, and an illustration of one is shown in Figure 69.

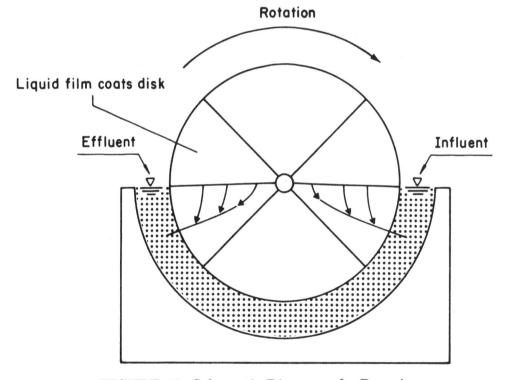

FIGURE 69. Schematic Diagram of a Rotating
Biological Contactor.

These units are too new for any rational basis of design to have been adopted. They are designed on an empirical basis with a hydraulic loading rate of approximately 50 ℓ/m^2 − day and a BOD loading rate of 0.2 kg/m^2 – day. Structural support of the plastic media must be sufficient to carry the additional weight of the biomass in rotation. If designed, constructed, and operated well, the units work efficiently, even with large variations in BOD concentration, with little energy input.

4. Activated Sludge

The conventional activated sludge process takes place in an aerated reactor. Here, again, the reaction in Equation 1.3 is carried out in an engineered system. The incoming wastewater is mixed with heterotrophs, usually provided in an excess concentration, and aeration provides an excess concentration of dissolved oxygen. The BOD concentration, therefore, becomes the limiting reactant in Equation 1.3. The heterotrophic microorganisms utilize the BOD as substrate, and when the substrate has been depleted, the reactor contents must be discharged from the reactor and the solids into a final sedimentation tank, called a secondary clarifier. Here the 'solid' sludge phase, consisting of the microorganisms, is separated from the 'purified' liquid and then recyled at a very high concentration back to the aerator or sent from the system for further treatment and disposal.

Often the tank is constructed in a long, narrow shape so that the tank approaches a plug flow reactor in which 'no' mixing occurs. Of course, some mixing does occur since the microorganisms, DO, and substrate must be mixed. But we say that a plug flow reactor is approached, so that the theoretical plug flow model can be utilized to simulate the system. We will use the other theoretical model to simulate the aeration tank, a completely mixed reactor. In this idealized reactor, 'instant and complete' mixing occurs so that the incoming BOD concentration is immediately diluted by the total volume of the reactor. The latter reactor should be more square-shaped or round in order to facilitate mixing. Both reactor types have advantages and disadvantages, but it must be remembered that neither ideal model can be achieved in reality. Figure 70 illustrates the notation to be used for designing activated sludge reactors on the basis of mass balance concepts.

Because many activated sludge plants were built on an empirical basis prior to the development of reactor theory, and because no ideal reactor behavior can be achieved in the real world, even modern activated sludge systems are still often designed empirically. The following specifications are usually employed for empirical design purposes:

Recently, much work has been carried out in order to apply basic reactor design kinetic procedures to the design and operation of

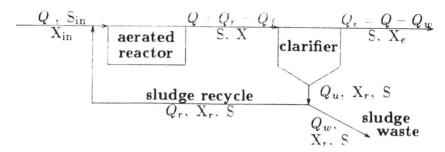

FIGURE 70. Schematic Diagram of Activated Sludge Secondary.

a. Hydraulic detention time of 6 hours,

b. Volumetric loading of 50 kg BOD per 100 m^3 of aeration capacity per day,

c. BOD loading rate of 0.4-0.5 kg BOD per kg biomass per day.

activated sludge aeration tanks. As a result, a better understanding of the process and better operation of the reactors has been achieved. These calculations utilize mass balance concepts presented in Chapter 4. Homework problems in this chapter apply these concepts illustrated in Example 3 as well as the empirical design concepts illustrated in Example 4.

EXAMPLE 3. Determine the theoretical hydraulic detention time and volume of a completely mixed reactor with recycle to be used in an an activated sludge treatment plant operating at steady state if the following conditions and constants for the wastewater have been determined:

X_{in} and $k_d \sim 0$ Efficiency $= 92\%$
$X = 2000$ mg/ℓ $BOD_{in} = 300$ mg/ℓ
$k_0 = 0.08$ hr^{-1} $Q + Q_r = 0.1$ m^3/sec
$K_M = 75$mg/ℓ

SOLUTION: Using Equation 5.47 at steady state and a known concentration of microorganisms,

$$0 = S_{in} - S - \frac{k_0 S X \Theta}{K_M + S}$$

$$0 = 300 \ \tfrac{mg}{\ell} - 24 \ \tfrac{mg}{\ell} - \frac{(0.08 \ hr^{-1})(24 \ \tfrac{mg}{\ell})(2000 \ \tfrac{mg}{\ell}) \Theta}{(75 \ \tfrac{mg}{\ell}) + (24 \ \tfrac{mg}{\ell})}$$

And, rearranging yields the following result:

168

$$\Theta = \frac{\left(276 \text{ mg}/\ell\right)\left(99 \text{ mg}/\ell\right)}{\left(3840 \text{ mg}^2/\ell^2\right)(\text{hr}^{-1})}$$

$$\Theta = 7.1 \text{ hr}$$

From Equation 5.10, the tank volume can be calculated,

$$V = Q\Theta$$
$$= \left(0.1 \frac{\text{m}^3}{\text{sec}}\right)(7.1 \text{ hr})\left(\frac{3600 \text{ sec}}{\text{hr}}\right)$$
$$= 2560 \text{ m}^3$$

EXAMPLE 4. Calculate the required aeration tank volume for an activated sludge treatment plant on the bases of the empirical design factors. The anticipated waste volumetric flow rate is 0.044 m³/sec and the BOD concentration is 250 mg/ℓ. Assume that the concentration of microorganisms in the reactor is kept at 2000 mg/ℓ.

SOLUTION:

ON THE BASIS OF HYDRAULIC DETENTION TIME:

For a 6 hour detention time -

$$V = Qt$$
$$= \left(0.044 \frac{\text{m}^3}{\text{sec}}\right)(6 \text{ hours})\left(3600 \frac{\text{sec}}{\text{hours}}\right)$$
$$= 950 \text{ m}^3$$

ON THE BASIS OF VOLUMETRIC LOADING:

For 50 kg BOD/ 100 m³ of aeration capacity per day

$$\text{BOD loading rate} = Q \times \text{BOD}$$
$$= \left(\frac{0.044 \text{ m}^3}{\text{sec}}\right)\left(\frac{0.250 \text{ kg}}{\text{m}^3}\right)\left(\frac{86,400 \text{ sec}}{\text{day}}\right)$$
$$= 950 \text{ kg/day}$$

169

$$V = \left(\frac{950 \text{ kg}}{\text{day}}\right)\left(\frac{100 \text{ m}^3 - \text{day}}{50 \text{ kg}}\right)$$

$$= 1900 \text{ m}^3$$

ON THE BASIS OF FOOD:MICROORGANISM:

For a food:microorganism ratio of 0.4 kg BOD/ kg microorganisms per day

BOD loading = 950 kg/day (as above) and with 2000 mg/ℓ microorganisms,

$$V = \frac{\left(\frac{950 \text{ kg BOD}}{\text{day}}\right)}{\left(\frac{2 \text{ kg microorganisms}}{\text{m}^3}\right)\left(\frac{0.4 \text{ kg BOD}}{\text{kg microorganisms} - \text{day}}\right)}$$

$$= 1188 \text{ m}^3$$

Since the largest design volume resulted from the volumetric loading design basis calculation, 1900 m^3 will be the stipulated aeration tank volume.

The great advantage of the activated sludge design, however, is that the liquid can pass through the system and be discharged from the secondary system. The microorganisms, after separating out in the secondary clarifier, can to a large extent, be recycled back to the reactor. The hydraulic detention time and the solid (cell) detention times, therefore, are VERY different. The hydraulic detention time is on the order of hours, since it makes a single pass through the secondary stage of the treatment plant. The cell detention time, on the other hand, is on the order of days, since the cells are recycled around the secondary stage over and over again.

The **mean cell detention time**, or sludge age, is defined as the biomass in the reactor divided by the biomass generation. Under steady-state conditions, the rate of biomass generation equals the rate at which biomass is discharged from the system. Using the nomenclature illustrated in Figure 70, the biomass in the reactor is defined by VX in the numerator, and the rate at which biomass is discharged from the system is the sum of solids discharged from the sludge waste line and solids overflowing the secondary clarifier weir into the effluent.

170

$$\Theta_C = \frac{VX}{Q_w X_r - Q_e X_e} \tag{7.2}$$

where Θ_C is the CELL detention time. If, as before, we assume that X_e approaches zero, the expression simplifies to the following:

$$\Theta_C = \frac{VX}{Q_w X_r}. \tag{7.3}$$

Equation 7.3 indicates that decreasing the rate of sludge wasting increases the sludge age or mean cell detention time. Since $Q_w = Q_u - Q_r$, decreasing the sludge wasting rate requires that the sludge recycle rate be increased if X_r remains constant. Therefore, by increasing the sludge recycle and decreasing the sludge waste, the solids detention time, Θ_c is increased. The aeration reactor, therefore, can be designed to to hold the wastewater long enough for the large volume of liquid wastewater to be degraded to the required efficiency. The secondary clarifier, however, must be designed to accommodate the larger concentration of microorganisms present due to the larger recycle ratio. Usually, any plant failure is due to an inadequate clarifier design and operation.

C. Secondary Clarifier Design

A secondary clarifier is a type of sedimentation tank. Unlike the Type I model used to design rectangular primary sedimentation tanks in Chapter 5, Stoke's Law can not be used to design secondary clarifiers carrying out Type II sedimentation. Type II sedimentation is characterized by flocculating or coagulating particles settling out from dilute suspension. These flocculating particles continue to gather mass as they flocculate and descend through the depth of the clarifier. The particle mass increases as the particle descends causing an increasing effect due to gravitational forces on the particle. A terminal settling velocity is never achieved by these particles, so that they do not follow a linear trajectory through the tank depth as do Type I settling particles. Type II sedimentation, therefore, is more complex. Usually this type of settling is referred to as clarification, and it is imposed to separate viable secondary microbial cells from the treated effluent. Often a circular tank with a conically-shaped tank bottom is designed for this purpose, and Figure 71 illustrates such a tank with the curved trajectory of Type II settling particles. Enhancing the curved shape of the trajectory is the effect of the decreasing horizontal advective velocity as the water moves from the center to the perimeter of the tank. If

171

the clarifier is designed to provide sufficient surface area, flocculating secondary mixed liquors can be introduced into the center of the tank and the sludge particles, as they descend, become so concentrated that they 'sweep' out even small particles which would not otherwise settle out and be removed from suspension.

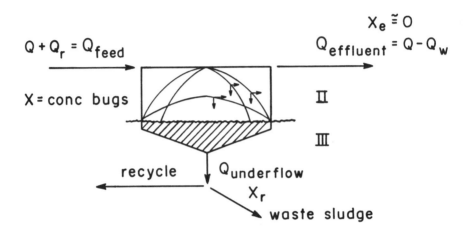

FIGURE 71. Type II Sedimentation in Secondary Clarifier.

An empirical design approach is used for a Type II sedimentation tank, or or secondary clarifier, for no rational basis exists for this purpose. Perhaps a common cause of process failure in the secondary stage of wastewater treatment is due to the lack of adequate secondary clarification. Example 5 illustrates the approach used for secondary clarifier design. For this purpose, a laboratory settling analysis must be carried out in a cylinder fitted with multiple ports, from which suspension is periodically withdrawn.

EXAMPLE 5. A laboratory settling analysis is conducted on a flocculating biological sludge from an existing nearby wastewater treatment plant thought to be similar to sludge to be generated in a new community. The laboratory cylinder used for analysis is 2.2 meters in height and contains ports at 0.6 m, 1.2 m, and 1.8 m in depth. The following data are collected:

% Suspended Solids Removed at Depths Specified

Time	0.6 m	1.2 m	1.8 m
5	41	20	16
10	55	34	31
20	61	45	39
40	70	58	54
60	78	62	58
90	86	70	63
120	90	76	72
150	95	83	76
180	98	90	85

Design a clarifier to remove 95% solids for the mixed liquor influent which enters the clarifier at 3000 mg/ℓ and a clarifier feed flow rate of 0.1 m^3/sec.

SOLUTION: First a plot of isoconcentration values is drawn up as shown below in Figure 72. Then calculations are carried out to determine the total removal at various detention times according to the following formula:

$$R_{\text{time interval 1}} = (\text{lowest \% removal at that interval})$$

$$+ \frac{(\text{midheight to next isoconcentration line})}{(\text{total height})}(\text{isoconcentration}_2 - \text{isoconcentration}_1)$$

$$+ \frac{(\text{midheight to next isoconcentration line})}{(\text{total height})}(\text{isoconcentration}_3 - \text{isoconcentration}_2)+$$

$$+ \ldots \ldots \text{etc.}$$

A typical calculation follows:

$$R_{58\text{ min}} = 50 + \left(\frac{1.3}{1.8}\right)(60 - 50) + \left(\frac{0.85}{1.8}\right)(70 - 60)$$
$$+ \left(\frac{0.45}{1.8}\right)(80 - 70) + \left(\frac{0.2}{1.8}\right)(90 - 80)$$
$$+ \left(\frac{0.14}{1.8}\right)(95 - 90) + \left(\frac{0.08}{1.8}\right)(100 - 95)$$
$$= 65.6\%$$

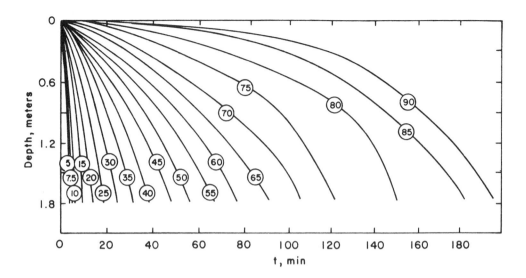

FIGURE 72. Iso-concentration Plot for Type II Settling Analysis.

and, similarly,

$$R_{80 \text{ min}} = 72\%$$

$$R_{120 \text{ min}} = 82\%$$

$$R_{150 \text{ min}} = 87\%$$

And calculating comparable overflow rates as follows:

$$q = \left(\frac{\text{Height}}{\text{time}}\right)(1440 \text{ min/day})$$

% removal	height/time, m/min	q, m^3/m^2 − day
70	1.5/100	22
80	1.1/140	16
90	1.2/180	12

By plotting Overflow Rate and Detention Time vs % Removal as shown in Figure 73, the necessary detention time and overflow rate can be obtained to design the tank.

Once the necessary detention time and surface overflow rate are determined from Figure 73, then the design is carried out as before. Figure 73 indicates that a surface overflow rate of 9 m^3/m^2 − day would be appropriate for the clarifier. Use Equation 6.12 to determine the necessary surface area.

174

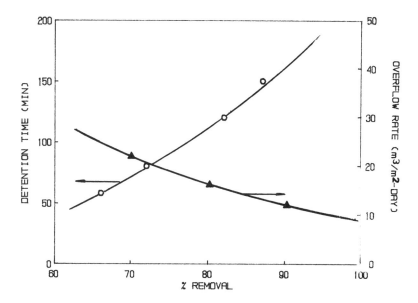

FIGURE 73. Design Determination for Type II Sedimentation Tank.

would be appropriate for the clarifier. Use Equation 6.12 to determine
the necessary surface area.

$$A_s = \frac{Q}{q}$$

$$= \frac{\left(0.1 \ \frac{m^3}{sec}\right)\left(86,400 \ \frac{sec}{day}\right)}{\left(10.5 \ m^3/m^2 - day\right)}$$

$$= 823 \ m^2$$

In secondary clarifiers, both effluent clarification and thickening occur. The clarification is caused by Type II (flocculating) sedimentation
phenomena, while thickening is caused by Type III (hindered) sedimentation. If a secondary clarifier is performing adequately, then clarified
effluent and thickened sludge are produced as shown in Figure 71.

A balance of volumetric flow rates around the secondary sedimentation tank yields the following expression:

A balance of volumetric flow rates around the secondary sedimentation tank yields the following expression:

$$Q_e = Q_f - Q_u \qquad (7.4)$$

where, Q_e is the volumetric flow of the effluent discharge,
Q_f is the volumetric flow in the feed line $(Q + Q_r)$,
and, Q_u is the volumetric flow rate of the underflow,
$(Q_r + Q_w)$.
All units are volume / time, usually m^3/sec.

Therefore, if Q_e is to be large, Q_u must be small. Consequently, if sufficient microbial concentrations are to be recycled back to the aeration tank, X_r must be very high. For a complete mix reactor with recycle, the solids loading is determined by the recycled solids term in the mass balance, $Q_r X_r$. A mass balance for solids at the juncture of the solids recycle line and the influent flow to the secondary aerator may be expressed as follows:

$$QX_{in} + Q_r X_r = (Q + Q_r)X \qquad (7.5)$$

The first term in Equation 7.5 becomes zero if X_{in} is assumed to be zero, as was assumed before in Chapter 5 when the mass balance was used to carry out reactor design. The two remaining terms can then be used to calculate the necessary recycle solids concentration X_r, if Q, Q_r, and X are known.

REFERENCES

[1] W. W. Eckenfelder, *Industrial Water Pollution Control*, McGraw Hill Book Company, New York, 1966.

For an overall view of secondary treatment, the same references as Chapter 3 are recommended.

HOMEWORK PROBLEMS

1. Rank the four types of secondary treatment systems with respect to efficiency and hydraulic capacity.

2. Which reactor types use 'fixed film' microorganisms and

which use 'free suspended' microorganisms?

3. For a low rate trickling filter reactor, determine the volume and surface area on the basis of empirical design factors for a wastewater with $Q = 0.04$ m^3/sec and BOD $= 250$ mg/ℓ.

4. The following laboratory tests have been carried out on a wastewater to determine the expected BOD reduction in a trickling filter. The incoming BOD $= 200$ mg/ℓ and the state regulatory agency limits the hydraulic loading rate to 4 m^3/m^2 – day. Calculate the percent BOD reduction.

	Fraction S remaining, S_{ef}/S_{in}		
depth, m	$q = 4.3$	$= 8.6$	$= 16.8$
	m^3/m^2 – day		
0.75	0.5	0.6	0.7
1.50	0.3	0.4	0.6
2.25	0.2	0.3	0.4
3.00	0.1	0.2	0.3

5. A completely mixed reactor without recycle is operated with $Q = 0.06$ m^3/sec and an incoming BOD $= 300$ mg/ℓ. The effluent BOD $= 30$ mg/ℓ. The following constants have been determined for this particular wastewater:

$k_0 = 0.4$ hr^{-1}
$K_M = 100$ mg/ℓ
$k_d = 0.005$ hr^{-1}
$Y = 0.5$

If X $= 1500$ mg/ℓ, calculate the reactor volume required, the theoretical detention time, and the BOD loading rate.

6. A completely mixed activated sludge tank with recycle is operated with $Q = 0.04$ m^3/sec and an incoming BOD $= 350$ mg/ℓ. The recycle ratio, $\mathcal{R}$, is 0.4, $Q_w = 10$ m^3/hr, and the recycle sludge concentration, X$_r$, is 7000 mg/ℓ. If the constants for this wastewater are the same as for Problem 5 and the reactor biomass concentration, X, is 1500 mg/ℓ, calculate the aerator tank volume required for a 90% reduction in substrate concentration, the BOD loading rate per 100 m^3 of tank volume, and the food: microorganism ratio.

7. Design an activated sludge aeration reactor for the wastewater described in Problem 6 on the bases of empirical design factors.

8. Design a circular secondary clarifier for Type II sedimentation to remove 90% of the clarifier feed concentration of 1800 mg/ℓ and a feed flow of 0.2 m^3/sec if the following laboratory data have been collected from a test described in Example 5:

% Suspended Solids Removed at Depths Specified

Time	0.6 m	1.2 m	1.8 m
10	10	6	3
20	28	18	14
30	40	30	23
40	56	36	32
50	66	46	36
60	74	56	48
80	84	68	55
100	91	75	63

COMPUTER PROBLEMS

Using the values for h, K, and n as in Example 2, calculate the necessary hydraulic loading rates for treatment efficiencies varying from 85% - 95% in increments of 1%. Which cases violate the most and least conservative hydraulic loading rate limits? How could h or A_s be varied in order to permit the system to operate within these hydraulic application limits? (See Table 9)

OR

Using values for the wastewater in Example 3, determine whether a zero order, first order, or Michaelis-Menten model would require the smallest reactor for 90% efficiency. Use the canned RKF45 program to integrate two differential equations simultaneously for the substrate and bugs.

Chapter 8

SLUDGE TREATMENT AND DISPOSAL

A. Treatment Processes

The more common sludge treatment processes can be grouped into the following major categories: (1) thickening, (2) digestion, (3) dewatering, and (4) drying. Sludge thickening and dewatering result in sludge-volume reduction, while sludge digestion has the major function of cell solids destruction and stabilization. Heat drying and combustion can accomplish both volume reduction and solids destruction. Numerous combinations of two or more of the treatment processes shown in Table 10 have been designed and operated successfully in order to achieve a given degree of sludge treatment.

1. Sludge Thickening Process

Sludge thickening is one of the most widely used and significant wastewater treatment processes available. Its purpose is to separate the sludge from the suspending liquid in a highly concentrated form, because the cost effectiveness of every subsequent sludge treatment, utilization, and disposal process depends on the suspended solids concentration of the sludge slurry. Gravity settling basins are the most often employed for the purpose of thickening the sludge slurry. Often Equation 8.1 is used to estimate the reduction in sludge volume, V, with concentration change,

179

TABLE 10
Summary of Sludge Treatment, Disposal, and Utilization Processes

A. Sludge Treatment Processes
1. Thickening
    a. Secondary Clarifier Thickening
    b. Separate Thickening Tank
        i. Gravity Thickening
        ii. Flotation
2. Digestion
    a. Aerobic
    b. Anaerobic
3. Dewatering
    a. Drying Beds
    b. Vacuum Filtration
    c. Pressure Filtration
    d. Centrifugation
4. Drying and Combustion
    a. Heat Drying
    b. Incineration
        i. Multiple Hearth
        ii. Fluidized Beds
    c. Wet Air Oxidation

B. Final Sludge Disposal Methods
1. Landfill
2. Lagoons
3. Ocean Dumping

C. Sludge Utilization Techniques
1. Biological Sludges
    a. Animal Feed Supplements
    b. Auxiliary Fuel
    c. Fertilizer and Soil Conditioner
2. Chemical Sludges
    a. Resource Recovery
    b. Complementary Chemical Usage
    c. Refinement for Marketing
3. Physically Inert Sludges
    a. Construction Filler
    b. Soil Conditioning

180

$$\frac{P_2}{P_1} = \frac{V_1}{V_2} \qquad (8.1)$$

where P is the percent solids by weight and V is the sludge volume. The subscripts correspond to conditions before and after the thickening process.

The thickening process can be carried out in separate thickening tanks or in conventional secondary clarifiers. A more concentrated sludge is obtained, generally, with separate thickeners, however. Mechanical and dissolved air flotation thickeners are less frequently used to thicken sludges from various sources in wastewater treatment plants. The sludge flotation process will not be discussed at all here, and the following discussion will present only gravity thickening.

In the thickening zone of the clarifier, sludge particles can not subside freely. Instead, the concentration of sludge is so high that neighboring particles hinder the free fall of any sludge particle. The principal design parameter for a sedimentation tank is the surface area. If a large surface area is provided, more solids can be loaded into the tank at a given Q_f before solids hindering occurs at some limiting concentration, X_L. The area provided must be large enough so that solids loading does not exceed the rate at which solids can reach the bottom. In Type III sedimentation theory, the term solids flux is used. The solids flux, N, is defined as mass of sludge solids passing through a horizontal plane in the tank, defined by the surface area per unit time,

$$N \;=\; \mathrm{mass/m^2 - time}$$

Solids flux due to gravity sedimentation is defined by the concentration times the subsidence rate. The subsidence rate is a Type III sedimentation rate, v.

$$N_{\mathrm{batch}} \;=\; Xv \qquad (8.2)$$

In a continuous flow sedimentation tank or thickener, solids subside due to the removal of the sludge at the bottom of the tank in the recycle line. Here, the underflow rate, u' is defined as Q_u/A,

$$N_{\mathrm{continuous\ flow}} \;=\; Xu' \qquad (8.3)$$

The total solids flux, then is the solids passing through the tank due to both gravity subsidence and removal in the recycle line.

$$N_{\mathrm{total}} \;=\; X(v + u') \qquad (8.4)$$

In order to increase N, the particles can be flocculated so that v increases, or Q_u can be increased in order to increase u. But if a solids mass balance is carried out around the clarifier,

$$Q_f X_f = Q_u X_r + Q_e X_e \tag{8.5}$$

and if the second term on the right hand side approaches zero when X_e approaches zero, then Equation 8.7 can be rearranged as follows:

$$Q_u = \frac{Q_f X_f}{X_r}$$

and since $Q_f X_f$ can not be increased or decreased at will, in order to increase Q_u, X_r must decrease. Therefore, if Q_u is increased indefinitely, the concentration of solids returned to the aeration tank may decrease to an insufficient concentration for carrying out biodegradation in the reactor. In order to increase N, then, polymer must be added to increase v, the design area must be increased in order to pass more solids to the bottom of the sludge thickening tank , or Q_u must be increased at the risk of diminishing the concentration of solids returned to the aeration tank.

The following section will present the design and operation procedure of sludge thickening tanks on the basis of batch solids flux determination only. Such a design and operation approach is a conservative one, for the additional solids flux due to the continuous removal in the sludge underflow line can only increase the total solids flux. The alternative approach of using total solids flux introduces an unnecessary complication for our purpose.

In a batch thickening test, a very concentrated slurry of sludge is introduced into a batch reactor, and at $t = 0$, the slurry is uniformly mixed and then allowed to remain under quiescent conditions. Soon a solids/liquid interface is formed which can be monitored as it subsides toward the bottom of the reactor. A subsidence vs time curve can be determined, as shown in Figure 74, and a corresponding subsidence velocity determined from the straight line portion of the curve.

If such an experimental determination is carried out at a variety of sludge concentrations, the relationship of subsidence velocity vs. sludge concentration can be determined. And, since batch solids flux is $N = X \times v$, the batch flux can be determined from that data. Both results are shown in Figures 75 and 76.

Furthermore, in a continuous flow thickener, the solids flux due to sludge removal from the bottom is $N = X_r \times u'$. Then graphically, u' can be determined by $u' = - dN/dX_r$. Since the data from the batch

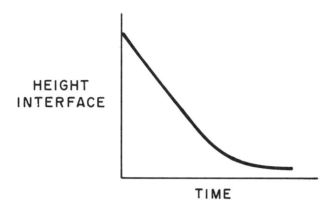

FIGURE 74. Subsidence vs. Time.

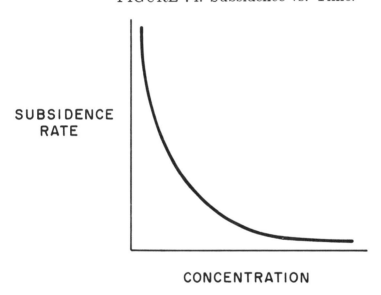

FIGURE 75. Subsidence Rate vs. Concentration.

analysis is plotted as N vs. X, a tangent from X_r, the desired thickened solids concentration, to the batch flux curve at or near the point of inflection will define u', the rate at which solids must be removed in the underflow line. The y-intercept defines the limiting flux, N_L, and the x-intercept defines the recycle solids concentration, X_r. An example calculation follows.

EXAMPLE 1. What size tank is required for thickening a sludge from 7000 mg/ℓ (0.7% solids) to 2% solids (20,000 mg/ℓ) if the volumetric flow rate into the tank is 0.0132 m³/sec?

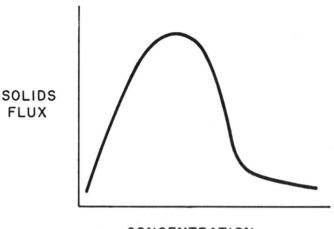

CONCENTRATION

FIGURE 76. Batch Flux Curve.

SOLUTION: As previously described, a subsidence velocity analysis must be carried out in order to determine the relationship between sludge concentration and subsidence velocity. From this data, the relationship between solids flux and solids concentration can be determined. See plot below.

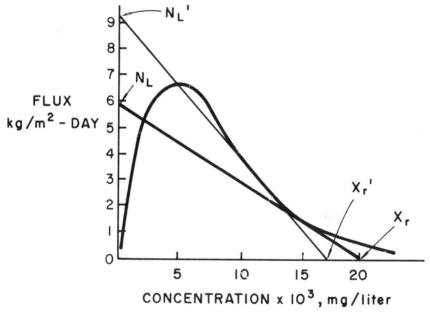

A tangent to the batch flux curve is shown by a heavy line from the desired recycle solids concentration of 20,000 mg/ℓ to the y-intercept. The resulting limiting solids flux, N_L, is 5.93 kg solids /m^2 − day. The solids loading, $Q_f X_f$ is calculated as follows:

184

$$Q_f X_f = \left(\frac{0.0132 \text{ m}^3}{\text{sec}}\right)\left(\frac{7 \text{ kg}}{\text{m}^3}\right)\left(\frac{86,400 \text{ sec}}{\text{day}}\right)$$

$$Q_f X_f = \frac{7983 \text{ kg}}{\text{day}}$$

A dimensional analysis demonstrates that $A_s N_L = Q_f X_f$. From this relationship, the tank surface area can be calculated.

$$A_s = \frac{Q_f X_f}{N_L}$$

$$A_s = \frac{\left[\dfrac{7983 \text{ kg}}{\text{day}}\right]}{\left[\dfrac{5.93 \text{ kg}}{\text{m}^2-\text{day}}\right]}$$

$$A_s = 1346 \text{ m}^2$$

Let us assume that a sludge thickener of this dimension is constructed. Then, if the solids loading varied, operation of the sludge removal could be varied. For example, if the solids loading decreased, the efficiency of the thickening process would only increase. Conversely, if the solids loading increased, eventually the limiting flux for the design would be exceeded and sludge solids would rise up and overflow the effluent weir into the clarified effluent to be discharged. For example, if the solids loading increased to 12180 kg/day, 4197 kg of solids would be lost per day (12180 kg/day - 7983 kg/day). Instead, calculate a new increased solids flux, N_L' for the batch flux curve and determine the corresponding decrease in X_r by defining a new operating line.

$$N_L' = \frac{Q_f X_f}{A_s}$$

$$N_L' = \frac{\left[\dfrac{12180 \text{ kg solids}}{\text{day}}\right]}{1346 \text{ m}^2}$$

$$N_L' = 9.05 \frac{\text{kg}}{\text{m}^2 - \text{day}}$$

A new, lighter-lined tangent from the new limiting flux, N_L', to the previous batch flux curve indicates that the new y-intercept establishes the new recycle solids concentration, X_r', at a value of 17,500 mg/ℓ. Therefore, the thickening efficiency is diminished, but the solids remain in the system for recycling or wastage. The solids are thereby prevented from overflowing the effluent weir and contaminating the discharge.

---

2. Sludge Digestion

Sludges such as those from biological waste treatment processes that have a high organic content can be subjected to digestion processes which serve to break down the cellular components and thereby decrease the total mass. This treatment can be accomplished aerobically, requiring oxygen and small tank sizes; the process, however, is generally carried out anaerobically, a slower procedure requiring a longer detention time. Anaerobic treatment eliminates the expense of the oxygen requirement and, furthermore, it produces methane—a useable end product which can serve as an energy source at various points in the wastewater treatment plant.

AEROBIC SLUDGE DIGESTION. Aerobic digestion is used to stabilize waste sludge solids by long-term aeration. The most common application of aerobic digestion is in reducing the cellular bulk of waste activated sludge. The excess biological sludge is aerated in the absence of any external organic carbon source and utilizes Equation 1.3. Under these endogenous respiration conditions, a fraction of the microorganisms will lyse and release their cellular constituents back into solution. The released materials serve as substrate for the remaining microorganisms which, under the aerobic conditions, will convert the released organic material to new cell mass and carbon dioxide and water in accordance with Equation 1.3. The overall reaction under these conditions is a net loss in cell mass, equivalent to the fraction of the cellular mass which has been oxidized. The cells are actually in the endogenous respiration phase. The longer the cell residence time, the greater will be the amount of cell destruction achieved, provided that sufficient oxygen is supplied.

ANAEROBIC SLUDGE DIGESTION. Anaerobic digestion is a widely-used process for the treatment of primary sludges and excess biological sludges, particularly from activated sludge and trickling filter operations. The process normally is conducted in a covered tank which is heated to maintain the sludge in the range of 35° C., the optimum temperature for mesophilic anaerobic bacteria. The biochemistry of the metabolic reaction taking place is in accordance with Equations 1.5 and 1.6, a two-step reaction wherein the sludge cells are converted first to organic acids by acid-forming bacteria and converted subsequently to methane and carbon dioxide by the methane-forming bacteria.

Tanks designed for this process normally are cylindrically shaped, with a cone-shaped bottom ranging in size to 30 meters or more in diameter and 15 meters in depth. The process can be conducted either under what is known as standard rate conditions (conventional digestion) in which stratification of the digester contents occur, as shown in Figure 77, or it can be conducted under high rate conditions in which staged digester tanks are in series, with the first mixed and the second stratified for separation, as shown in Figure 78.

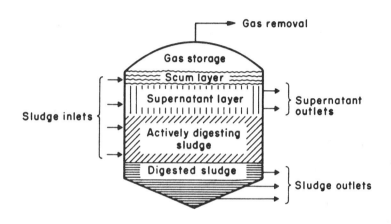

FIGURE 77. Schematic Diagram of Single Stage Digester.

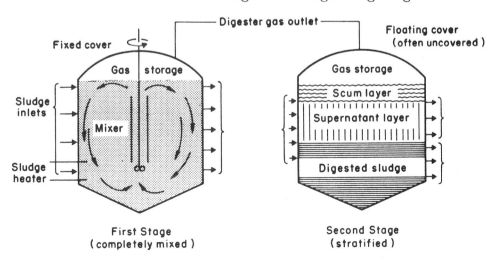

FIGURE 78. Schematic Diagram of High Rate Digestion.

The purpose of the mixing is to utilize the entire contents of the digester for active digestion; the second stage digester is provided under these conditions primarily for separation of the digested sludge and digester supernatant liquid. Additional digestion can be achieved also in the secondary digester, depending on the detention time of the sludge. Mixing is generally achieved in the first stage either by gas recirculation, pumping, or draft tube mixers. The high-rate digestion process differs from the conventional single-stage process in that the solids-loading rate is much greater and shorter hydraulic detention times are practical.

Sludge anaerobic digesters are designed on a number of different bases. One empirical basis is simply to calculate the percapita served by the treatment plant and design the total digester volume to accommodate the volume generated for the necessary detention time. Another basis to use is to provide sufficient volume to allow a given solids loading factor. *Ten States Standards* is a handbook for these empirical design approaches.

3. Sludge Dewatering

The methods available for sludge dewatering can be classified as either filtration-type methods (i.e., gravity, pressure, vacuum filtration) or centrifugation methods. The selection of a sludge-dewatering method depends on many factors. Generally, sand-drying beds have been used for smaller, domestic wastewater treatment operations where sufficient land usually is available at a reasonable cost. Dewatering of sludge organisms by this method requires pretreatment by anaerobic digestion in order to prevent odors and render the sludge dewaterable. Climatic conditions will also greatly affect this method. The method consists principally of pumping digested sludge uniformly onto sand strata, allowing the sludge to dry over a period of time, removing the dried cake (usually by manual means), and cleaning the bed, after which the process can be repeated. Digested sludge normally is applied to a sand-drying bed at a depth of 30 cm; sludge drying times of 1 to 3 months normally are allowed— depending on the sludge and climatic conditions.

Vacuum filtration has the inherent advantage of working well on a wide variety of sludges and producing a relatively dry filter cake which can be incinerated. The solids capture is very good by this method, although sludge preconditioning with chemical flocculants is normally required for effective operation. Vacuum filtration normally is conducted on a slowly rotating drum filter. The filtration drum consists of a series of cells which run the length of the drum and can be placed under vacuum as required. The drum is covered with a suitable filter medium (e.g., wool, cotton, synthetic filter cloth, steel mesh, etc.). The slurry to be dewatered is contained in a tank through which the filter drum rotates. The filter cake initially is formed on the filter medium as the drum rotates through the slurry tank and then is dried by liquid transfer to air drawn through the cake. At the end of the cycle, the dried cake is removed from the drum by a knife-edged scraper or equivalent device. A schematic drawing of a typical filtration system is shown in Figure 79.

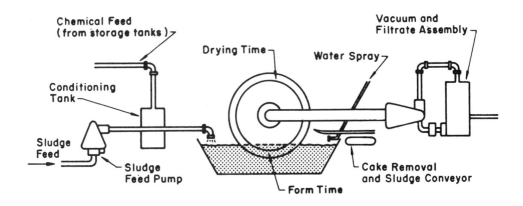

FIGURE 79. Schematic Diagram of a Vacuum Filter.

The operation of pressure filtration systems is similar to that of vacuum filtration in that a porous medium is used to separate the solids from the suspending liquid. As the solids are captured on the medium, they build up to form a cake through which the liquid is filtered. Sludge pumps provide the positive pressure to force the water through the cake—rather than pulling the water through by vacuum.

Filter-press operation classically has been on a batch basis and limited mostly to European installations. Recently, however, a number of filter presses have been developed which are capable of continuous operation and produce reasonably dry filter cakes. One such system is shown in Figure 80. This system consists of two moving belts running together at an acute angle, thus forming a narrow, tapering pressure space into which the sludge to be filtered is pumped. Dewatering takes place in the following two-stage sequence as the material moves continuously along the belt: (1) liquid stage which utilizes normal hydrostatic pressure, and (2) solid stage which requires application of low pressure.

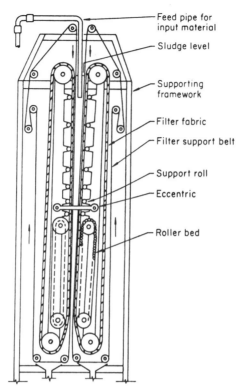

Feed pipe for
input material

Sludge level

Supporting
framework

Filter fabric

Filter support belt

Support roll

Eccentric

Roller bed

FIGURE 80. Schematic Diagram of a Moving Belt Filter Press.

Continuous-flow centrifuges are now utilized in many industrial and municipal sludge dewatering operations. Although the resulting centrate is generally of poorer quality than the corresponding filtrate would be from sludge filtration, the process has the advantage that the capital cost is low compared to other mechanical dewatering equipment, operating and maintenance costs are moderate, the unit is totally enclosed, odors are minimized, and operation is relatively simple. The most effective centrifuges for sludge dewatering are continuous-flow horizontal solid-bowl machines, as illustrated in Figure 81.

These machines are cylindrically shaped and fitted with a truncated cone section on one end. Sludge is continuously fed into a rotating bowl through a stationary pipe along the axis of rotation. As the solids enter the centrifuge, they are acted upon by the centrifugal force which deposits them against the wall of the bowl where they are scraped to the conical section by a screw conveyor and ultimately discharged. The liquid effluent (centrate) is discharged from effluent ports after passing the length of the centrifuge under centrifugal force.

190

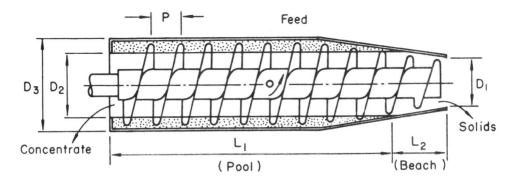

FIGURE 81. Schematic View of Continuous Flow
Solid Bowl Centrifuge.

4. Drying and Combustion

The incineration of sludge is becoming more and more widely practiced
as sufficient land for other treatment and disposal methods is more
difficult to obtain. Although the process is relatively expensive and
can create an air pollution problem if it is not properly operated and
controlled, sludge incineration can achieve both volume reduction and
solids stabilization. The major factors which affect the incineration
of sludge are the calorific or heat value of the solids and the residual
moisture content in the sludge cake.

The incineration process can be divided into two distinct operations: (1) drying and (2) combustion. The drying process entails the
raising of sludge temperature from the ambient to 100° C., evaporat-
ing the water from the sludge, and increasing the temperature of the
water vapor to the temperature of the exit gases from the incinera-
tor. The combustion process liberates heat from the chemical reaction
between the sludge fuel and oxygen. The organic material in sludge
provides the source of fuel for the incineration process. The amount of
auxiliary fuel required for the incineration of sludge can be determined
from a heat balance between the heat available from the sludge and
the heat required to: (1) evaporate the water from the sludge, (2) heat
the solids to ignition temperature, (3) heat the incoming air, and (4)
compensate for heat losses. If the organic content of a sludge is at least
40 to 50%, the dewatering process will generally provide material for a
self-sustaining combustion reaction, once the burning of auxiliary fuel
raises incineration temperatures to the ignition point.

Products of sludge incineration from municipal wastewater treatment plants are primarily ashes and stack gases. If the operating conditions are satisfactory, the ash is primarily sterile inorganic material,

191

while the composition of the stack gases will depend to a great extent on the type and operation of installed air cleaning devices. Results of mass spectrographic analyses of stack gases from municipal sludges produced under good operating conditions indicate that the gas is composed predominately of nitrogen (82%) with smaller percentages of carbon dioxide (9%) and oxygen (8%); hydrocarbons, oxides of nitrogen, hydrogen sulfide, and oxides of sulfur can be present under poorer operating conditions.

Incineration processes available for the burning of sludges include multiple hearth furnaces, rotary kiln furnaces, and fluidized bed reactors.

B. Sludge Disposal

In the final analysis, however, a residue will always result from any sludge treatment process. Thus, in order to achieve complete sludge treatment, this residual material ultimately must either be disposed of effectively or utilized in some fashion. As indicated in Table 10, Section B, sludge disposal techniques currently practiced consist principally of land or sea disposal techniques, depending on geographical location.

1. Landfilling

Sanitary landfills have been utilized as final disposal sites for virtually every kind of raw and treated sludge. If sludge is to be delivered to the landfill by truck rather than by pipeline, the economics will usually indicate that dewatering for volume reduction will result in justifiable savings. Operation of a landfill receiving sludge should be similar to one receiving solid wastes, i.e., the wastes should be deposited in a designated area, compacted in place with a tractor or roller, and covered with a 30 cm layer of soil. The sanitary landfill method is most suitable if it is used also for disposal of refuse and other solid wastes of the community. With daily coverage of deposited sludge, nuisance conditions such as odors and flies are minimized.

Sanitary landfills for sludge disposal have the inherent disadvantages of requiring large land areas, possibly polluting the groundwater, and having their operation affected by the weather. The use of sanitary landfills for sludge disposal from medium- to large-size urban cities is uncommon because land is too expensive or simply not available at these locations. In those cases where landfilling is practiced currently, more expensive but more compact sludge treatment-utilization techniques are being substituted at a rapid rate.

192

2. Lagooning

Lagooning is currently a widely practiced sludge disposal technique, particularly at industrial sites. Because lagooning is cheap and simple, it will continue to be used widely, provided that sufficient land is available at a reasonable cost. Various types of sludge lagooning operations exist. Lagoons have been used as thickeners, storage tanks, digestion tanks, drying lagoons (from which sludge is periodically removed), and permanent lagoons in which sludge remains for an indefinite period. If lagoons are constructed in areas where the soils are porous, the bottom must be lined to prevent groundwater contamination. Sludge from domestic wastewater treatment plants can generally be dewatered to about 55 to 60% moisture content in a two to three year period in a lagoon; a residual moisture content in the range of 60 to 70% is more common in most operations, however. Lagoons are subject to the same general disadvantages as landfills, i.e., large land areas are required, possible groundwater contamination can result and, in general, they are not feasible for medium or large size urban areas because land is either too expensive or simply not available. In addition, unpleasant odors frequently can result from lagoon operations.

3. Discharge to the sea

Ocean discharge of sludge has been practiced in past years by coastal cities. The practice has been less costly than alternative sludge disposal methods. Both pipeline discharge and barging have been employed successfully—with some barging distances totaling 400 miles round trip. The usual practice has been to anaerobically digest the sludge before ocean discharge in order to stabilize the sludge and thereby minimize problems associated with the disposal of raw sludge.

This method provides complete removal of sludge from the wastewater treatment plant and its environment. Removal of floatable materials before discharge and the selection of discharge sites to preclude any sludge deposits on beaches or other shoreline areas have been major considerations for this final sludge disposal technique. The suitability of this sludge disposal practice, however, has been questioned on numerous grounds, and this practice has been banned in the United States. In this regard, such questions have been raised concerning: (1) the impact of the sludge discharges on the aquatic environment, (2) the accumulation of bottom deposits, (3) the potential health hazard, and (4) the wasting of a land resource into the sea. Severely restricted US sludge discharge into the sea may be permitted at some future time, as many foreign countries now carry out this practice.

C. Sludge Utilization

The utilization of sludges from manufacturing and waste treatment processes has become more of a necessity in recent years. Such a trend will no doubt continue as disposal of increasing amounts of sludge becomes less feasible. 'Recycle' is a catchword which has been introduced and widely adopted by industry and the public. The brine wastes from analogous processes are more difficult both to dispose of and to utilize because the water content is so high; the large volume added by the water to the chemical content causes greater transportation costs for disposal or utilization. The number of potential uses for sludges and brines is limited only by the imagination and recycling costs. In the future, costs of recycling waste products may be added to the initial costs of manufactured goods. Table 10, for example, lists a number of utilization methods for various sludge types.

The South Lake Tahoe Wastewater Reclamation Project has pioneered the concepts of chemical recovery from sludges produced during the wastewater treatment process. In the total treatment scheme operated at that facility, lime is added to the stream in order to raise the pH so that calcium phosphate precipitation occurs and ammonia stripping can be carried out [1]. The lime is recovered from the calcium phosphate sludge by an elaborate but efficient recalcination process, so that the recovered lime may be added to the chemical feed of the next treatment cycle. Figure 82 illustrates the process. Depending on the initial calcium and alkalinity concentrations of the water and the recovery method selected, the quantity of 'recovered' lime actually may be greater than that added to the water and thus may completely eliminate the need for additional chemical. Regardless, utilization of such techniques, then, not only eliminates an otherwise necessary sludge disposal problem but also provides necessary chemical for the wastewater treatment process.

Recovery of heavy metals from sludge presently has been considered unsuccessful or economically unfavorable. A potential method utilizing high-temperature plasmas for the elemental recovery of all sludge constituents has been described by Eastland [2].

Laboratory scale studies have been conducted to show that closed-cycle food-wastes systems can be operated on a self-perpetuating basis [3, 4]. These systems involve the synthesis of algal sludges from wasteswaters, and the algae, in turn, serve as food for primary consumers.

It has been suggested that dried sludge could be used as an auxiliary fuel, since disposal techniques by incineration can be self-supporting if

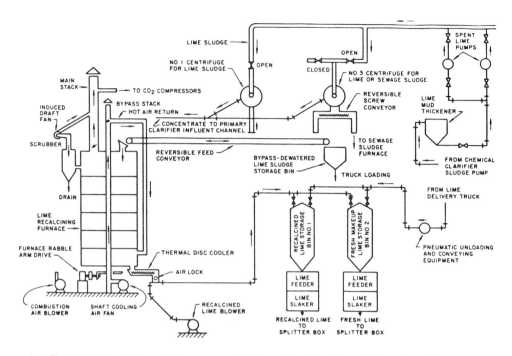

FIGURE 82. Flow Diagram of Lime Sludge Recalcination Process.

the organic content of the sludge is sufficient. In plants which dispose of sludge by incineration, instead of being wasted, the heat produced could be utilized as an energy source within the plant, e.g., to heat sludge digesters.

Sludge 'farming' has become a widely touted procedure for the utilization of waste sludge. The Chicago Metropolitan Sanitary District has operated such a facility very successfully [5, 6]. The cost of transporting the sludge to an outlying district has incurred increasing costs, however, and pipelines have been suggested for the transport of large volumes of sludge [7]. The advantageous aspects of this type of utilization are that the inorganic fraction of the sludge serves to condition the soil and the organic and nutrient fractions fertilize the soil. With such waste recycling, marginal or depleted fields can be made more productive. Certain aspects of this technique require additional investigation at this time, however, before widespread use of such methods can be endorsed; these considerations include both the fate of heavy metals and pathogenic microorganisms in the farming operation. Dried sludges from activated sludge operations in a number of cities (e.g., Milwaukee, Chicago, Winston-Salem, N.C.) have been bagged and sold as a form of fertilizer with varying degrees of success [8].

Other uses for waste sludge have been suggested or demonstrated. Gasification of sludge for producing low-molecular-weight hydrocarbons to serve as fuels is one example [9]. The slag produced from incinerated sludge has been studied as a road building and concrete manufacturing aggregate [10].

Many of these waste sludge utilizations are not economically or technologically feasible at the present time. As costs of ultimate disposal increase or when certain questionable disposal techniques are no longer permitted, however, these procedures will become more attractive. In addition, the ultimate need to recycle wastes, which currently are disposed of in some manner but could be utilized as a resource, will become more prevalent as resources become more scarce and expensive.

REFERENCES

[1] J. L. Steinfeld, *Reader's Digest,* p. 170 (November), 1978.
[2] B. J. Eastland, *J. Water Pollution Control,* **109**, 18 (1971).
[3] H. B. Gotaas, *et al., Scientific Monthly,* **79**, 368 (1954).
[4] G. L. Dugan, *et al., J. Water Pollution Control,* 44, 432, (1972).
[5] S. M. King, *Fert. Sol.,* **15**, 24 (May, 1971).
[6] R. B. Ogilvie, *Environment,* **1**, 1 (1971).
[7] T. L. Thompson, *Chem. Eng. Prog. Symp. Ser.,* **67**, 413 (1971).
[8] F. Styers, *Am. City,* **86**, 48 (1971).
[9] H. F. Feldman, *Chem. Eng. Prog.,* **67**, 51 (1971).
[10] Anonymous, *Environ. Sci. Technology,* 5, 197 (1971).

HOMEWORK

1. A sludge thickener tank receives a volumetric feed flow rate, $Q_f = 0.05$ m^3/sec. The feed solids concentration, X_f, is 5000 mg/ℓ (0.5% solids). A batch sludge subsidence analysis indicates that in order to thicken the sludge to 4% solids, limiting batch flux, N_L, is 10.0 kg solids/m^2 − day. Determine the necessary thickener tank surface area.

2. A sludge thickener must be designed for receiving 0.005 m^3/sec of feed flow, Q_f. The feed solids concentration of 5000 mg/ℓ (0.5% solids) must be thickened to 2% solids. Calculate

the thickener surface area if sludge thickening tests produce the following data:

v, m/hr	X, gm/ℓ
0.030	6.8
0.029	7.0
0.018	10.0
0.005	16.0
0.0004	35.0

3. After the thickener in Problem 2 is built, the feed flow rate actually averages 0.004 m^3 per sec. What is the maximum sludge concentration obtained at this solids loading?

COMPUTER PROBLEM

A function often used to describe the total flux from Equation 8.4 is as follows:

$$N_{total} = \frac{X_r \mathcal{R} Q}{A_s}$$

where

$$X_r = \frac{X(1 + \mathcal{R})}{\mathcal{R}}$$

Calculate N_{total} for the thickener in Example 1 if $\mathcal{R} = 0.3$. Set up a DO LOOP varying X from some value less than the incoming concentration to some concentration greater than the desired X_r. Compare the resulting N_{total} curve to the N_{batch} curve shown in Figure 76.

197

Chapter 9

TERTIARY WASTEWATER TREATMENT

A. Physical-Chemical Treatment

Utilization of physical and chemical processes for primary and secondary treatment of wastewaters in lieu of the more conventional combination of physical and biological processes is currently being practiced to some degree. This type of treatment is particularly advantageous for systems which experience appreciable fluctuations in pollutant concentrations or which contain pollutants that would be toxic to a biological wastewater treatment plant; also, the process is capable of removing appreciable concentrations of other pollutants in addition to the dissolved organic materials normally removed in secondary treatment. The treatment is accomplished in a series of unit operations which may be varied in any number or order. The physical-chemical treatment plant must either recover and/or regenerate the chemicals utilized to perform the treatment or, alternatively, new chemicals continually must be purchased. A flow diagram of a typical physical-chemical treatment plant is shown in Figure 83.

Preliminary treatment usually consists of screening, in order to remove the larger objects and could include a complete preliminary and primary treatment system as shown in Figure 54. But a major advantage of physical-chemical treatment schemes is that only a minimal amount of preliminary treatment is absolutely necessary. As shown in Figure 81, the first unit process following preliminary treatment in-

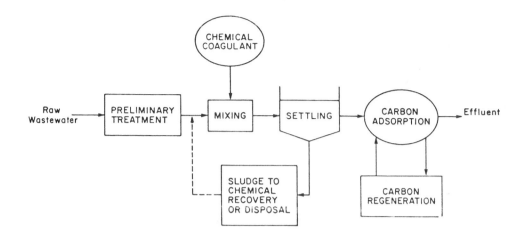

FIGURE 83. Flow Diagram of a Typical
Physical Chemical Treatment Plant.

volves chemical coagulation in a mixing tank. Inorganic metal ions are
best suited as coagulant chemicals because of their inherent dual role
of acting as both a chemical precipitant and a coagulant. Due to their
relatively low cost and general availability, the three most frequently
used metals are calcium, iron, and aluminum. After a period of rapid
mixing for approximately 1 min, followed by slow mixing for 15 to 30
min, the coagulated and precipitated impurities are effectively removed
by ordinary sedimentation.

The next stage, adsorption, utilizes tanks filled with carbon particles over which the liquid wastewater flows so that the dissolved carbonaceous pollution can be adsorbed onto the surface of the solid carbon particles in the filter bed. This adsorption process may continue
until all the adsorption sites on the carbon bed are occupied by the dissolved carbonaceous pollutant molecules from the wastewater. When
the filtration bed is saturated, the bed must be removed from the treatment stream and the carbon particles treated to remove the adsorbed
organic carbon molecules. By clearing the adsorption sites on the solid
carbon particles, we say that the carbon particles are regenerated and
thus can be utilized repeatedly. The following sections describe each
unit operation in more detail.

1. Chemical Coagulation

As described in Chapter 3, the inorganic salts of calcium, aluminum, and iron are the most suitable chemicals for chemical coagulation processes. Depending on the conditions, one of the following chemicals is generally selected: ferric chloride, ferric sulfate, sodium aluminate, aluminum sulfate (alum), or calcium oxide (lime). The necessary chemical dose can best be determined experimentally on a laboratory scale prior to treatment as illustrated in Example 3 of Chapter 3. If the chemical reaction for any precipitation reaction involved in the coagulation process is known, theoretical doses may be approximated from the stoichiometry of the precipitation reaction. The requisite concentration of metal-ion coagulant is influenced by such parameters as pH, alkalinity, temperature, mixing, and concentration of dissolved and suspended species, and, for the above salts, normally is in the range of 150 to 400 mg/ℓ. The agglomeration of individual particles is based on concepts described in Chapter 3 and is carried out in a slowly mixed coagulation tank. The clarification following chemical addition and coagulation depends upon the separation of the solid suspended particles from the suspending liquid phase. The separation generally is achieved by sedimentation or upflow clarification techniques and thus the solid particles must be in an agglomerated state to facilitate the separation. Synthetic organic polyelectrolytes have been used successfully in conjunction with the inorganic metal-ion coagulant chemicals in order to enhance the separation process.

2. Dissolved Organic Adsorption

Granular activated carbon has long been utilized to remove color and odors from water. Adsorption of color- and odor-causing substances on the carbon surface takes place because the carbon particles possess unsatisfied lattice charges and the dissolved substances contain polar or ionic groups of opposite charge. The attachment of the undesirable organic substance takes place at 'active sites' on the surface of the carbon particles. This adsorption is dependent primarily on concentration of dissolved substances, carbon particle size and related surface area, and flow rate of the liquid to be treated. The relationships used to express the adsorption process are developed empirically. They can be expressed graphically as isotherms or expressed algebraically. They are developed to determine the optimum operating parameters for the adsorption process. See Example 1.

EXAMPLE 1. An industrial waste containing 150 mg/ℓ of toxic phenol must be treated by activated carbon adsorption to reach the acceptable discharge concentration of 1.0 mg/ℓ phenol. Adsorption dosage tests (similar to those illustrated in Figure 39) indicate that at the 1 mg/ℓ discharge concentration, 0.5 mg phenol is removed per mg of activated carbon. How much activated carbon will be required per day to treat a waste flow of 0.01 m³/sec?

SOLUTION:

Required Daily Phenol Removal
$$= \left(0.01 \tfrac{m^3}{sec}\right)\left(86,400 \tfrac{sec}{day}\right)\left(\tfrac{150\ mg - 1\ mg}{\ell}\right)\left(\tfrac{10^3 \ell}{m^3}\right)$$

$$= \left(128.7 \times 10^6 \tfrac{mg}{day}\right)\left(\tfrac{1\ kg}{10^6\ mg}\right)$$

$$= 128.7 \tfrac{kg}{day}$$

Required Daily Carbon
$$= \left(128.7 \tfrac{kg}{day}\right)\left(\tfrac{1\ mg\ carbon}{0.5\ mg\ phenol}\right)$$

$$= 257.5 \tfrac{kg}{day}$$

A thorough theoretical treatment of carbon adsorption filters was published by Weber and Smith [1]. A newer variation in carbon adsorption incorporates a countercurrent flow system, in which the wastewater flows in a forward direction and powdered carbon is pumped in a reverse direction. Such a flow configuration is called 'expanded-bed' design and, although it requires longer columns, it provides greater surface area and, correspondingly, increased adsorption capacity and greater operating efficiency. Weber [2] studied operational characteristics of the concurrent and countercurrent designs and found that the conventional concurrent flow design required considerable maintenance in order to produce comparable removal efficiency. Figure 84 illustrates conventional and expanded designs.

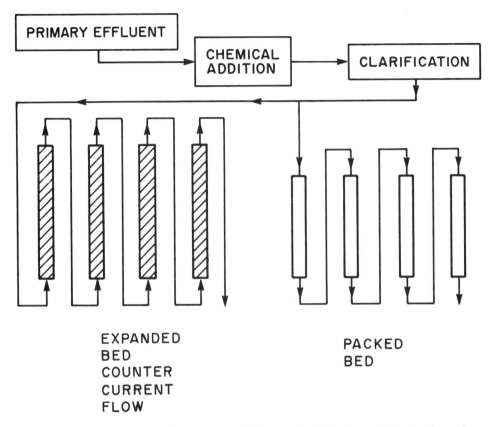

FIGURE 84. Flow Diagram of Expanded-Bed and Packed Bed
Carbon Adsorption Columns.

B. Tertiary and Advanced Wastewater Treatment

1. Phosphorus Removal

Quite commonly, wastewaters are overconcentrated with respect to phosphatic pollutants; i.e., there is a greater concentration of this nutrient than required soichiometrically for a carbon/phosphorus ratio of 106:1. Consequently, only a fraction of the phosphorus in domestic wastewater generally will be biologically removed by conventional secondary wastewater treatment systems. The two most common secondary biological treatment processes, trickling filter installations and activated sludge aeration facilities, have been reported to be capable of removing a maximum of 20% and 40% of the phosphorus concentrations, respectively, under ordinary operating conditions [3]. Therefore, additional treatment must be utilized if efficient removal of phosphorus is desired following secondary biological treatment. Efficient phosphate removal may be achieved by either microbiological means or chemical precipitation.

a. Microbiological Removal of Phosphorus

In accordance with Equations 1.3 and 1.4, microbiological removal of phosphorus is carried out by either heterotrophic or photo-autotrophic microorganisms. Because of the sunlight requirements, which might include provision of artificial illumination during periods of darkness and the use of relatively shallow tank or pond depths, as well as the generally difficult separation characteristics of algae, the use of autotrophic systems for the removal of phosphorus has been rather limited. The use of combined heterotrophic-autotrophic systems (i.e., lagoons or oxidation ponds) as described previously are widely used and, depending on their operating conditions, are capable of achieving appreciable percentages of phosphorus removal. Great attention, however, has focused on manipulating the operation of the activated sludge process in an attempt to achieve greater phosphorus removal percentages than the stoichiometric values. Ordinarily, the activated sludge microorganisms would contain approximately 2 to 3% of their volatile suspended solids concentration as phosphorus in accordance with the stiochiometry of their elemental composition of microbial protoplasm ($C_{106}H_{180}O_{45}N_{16}P_1$). By proper adjustment of such parameters as aeration rates, method of sludge recycle, and cation concentration of the wastewater, a number of activated sludge wastewater treatment plants have reported phosphate removals appreciably greater than 40% [4 − 7]. These removals far exceed any 106:1 stoichiometric carbon/phosphorus ratio and, at times, these treatment plants reportedly achieve a 90 to 95% phosphate removal, even in phosphate-rich wastewaters.

Such excessive phosphate removal by an activated sludge process has been termed 'luxury uptake of phosphorus', and much controversy has arisen over the mechanism by which this luxurious removal takes place. This phenomenon is often referred to as LUP, and studies [8] support the theory that luxury uptake of phosphorus is a biological mechanism operated by an 'active' transport system. Apparently, this mechanism is controlled by the cells' ability to expend energy for the purpose of transporting phosphate into the cell, despite higher concentrations inside the cell compared to the concentration in the suspending wastewater liquid. Figure 85 illustrates treatment plant designs which have been developed to exploit the luxury uptake of phosphorus.

Under aerobic conditions in which active transport can occur, the

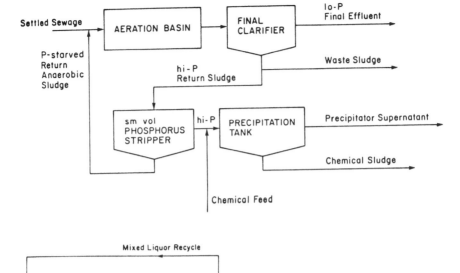

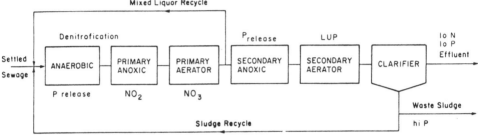

FIGURE 85. Schematic Diagrams of LUP Treatment.
(top) Phostrip Process; (bottom) Bardenflow Process.

cells can transport phosphate into the cells despite a higher concentration of phosphate inside; the subsequent release of phosphate from the biological cell occurs under anaerobic conditions, when the cell can not utilize the active transport system. In the Phostrip process, the sludge takes up phosphorus from the liquid phase in the aerator and releases it back to the liquid phase in the anaerobic phosphorus stripper. The phosphorus-stripped sludge is then recycled back to the aerator ready for more phosphorus uptake, while the phosphorus-rich suspending liquid is sent to a separate tank where precipitating chemicals are added to remove the phosphorus from the liquid phase. In the Bardenflow system, alternate reactors under aerobic and anaerobic (or anoxic, in which nitrates serve as electron acceptors in the oxidation process) conditions promote phosphorus uptake and nitrification (see below) followed by phosphorus release and de-nitrification. Phosphorus removal in the sludge is accomplished only by sludge wasting. The Bardenflow Process, however, removes both nitrogen and phosphorus.

b. Chemical Removal of Phosphorus

Specific metal ions are effective in precipitating phosphate from solution. Due to their relatively low cost and general availability, the three most frequently used metals for phosphate precipitation are calcium, iron, and aluminum. An examination of Table 11 reveals that all three of these metals form relatively insoluble precipitates with phosphate [9].

TABLE 11

PHOSPHATE PRECIPITATION REACTIONS AND CONSTANTS

REACTION	- $\log K$
$Fe^{+3} + PO_4^{-3} \rightarrow FePO_4$	23
$3\ Fe^{+2} + PO_4^{-3} \rightarrow Fe_3(PO_4)_2$	30
$Al^{+3} + PO_4^{-3} \rightarrow AlPO_4$	21
$Ca^{+2} + 2\ H_2PO_4^- \rightarrow Ca(H_2PO_4)_2$	1
$Ca^{+2} + HPO_4^{-2} \rightarrow CaHPO_4$	6
$10\ Ca^{+2} + 6\ PO_4^{-3} + 2\ OH^- \rightarrow Ca_{10}(OH)_2(PO_4)_6$	90

The tendency of aluminum and iron to hydrolyze in aqueous solution creates a competition between the hydroxide and phosphate ions for the coordination sphere of the metal. Thus the efficiency of phosphate removal is dependent upon the relative concentrations of these two anions in solution and is consequently pH-dependent. A decrease in pH or, more precisely, OH^-, favors precipitation of the metal phosphate. However, as the solubility of the metal phosphates increases with decreasing pH, an optimum pH exists for the removal of phosphate with metal-ion precipitants. When calcium is used as a precipitant, the competition for calcium is predominantly between the phosphate and carbonate anions and, again, phosphate removal is dependent upon the relative concentrations of the anions present, and upon pH. Hydroxylapatite, $Ca_{10}(OH)_2(PO_4)_6$, is the most stable calcium phosphate solid phase as shown in Table 11. In practice, therefore, the pH and alkalinity of a wastewater determine to a large degree the relative efficiency of phosphate removal by precipitation with metal ions.

The chemical precipitation of phosphorus is a two-stage process:

First, the phosphorus must be made insoluble through the use of metal ions in accordance with reactions such as those shown in Table 11; and, secondly, the precipitate must be removed from suspension by sedimentation. The second step thus requires additional chemical for conditioning or flocculating the precipitate for its subsequent separation. The addition of various synthetic organic polyelectrolytes, or the addition of excess metal ions serves identical purposes in bringing about agglomeration of the insoluble metal phosphate precipitates into a settleable sludge. Chemical removal of phosphorus generally has been designed as a separate third-stage treatment system, particularly when using lime. In the case of alum and ferric salts, successful removals have been achieved by chemical addition to existing secondary biological treatment units as well as to existing primary systems.

2. Nitrogen Removal

Appreciable consideration has been given to the removal of nitrogen from wastewaters because of its potential eutrophying characteristics and because of the oxygen-demanding effects of the more reduced forms of nitrogen (i.e., ammonia and organic nitrogen species). In addition, the presence of ammonia nitrogen in effluents has been shown to be toxic to aquatic life and an interference to the chlorination process. For these reasons, many states have promulgated effluent standards limiting the total concentration of ammonia-nitrogen and/or total nitrogen in wastewaters. Nitrogenous species can be removed from wastewater either by biological means or by physical-chemical methods.

a. Biological Removal of Nitrogen

The biological transformations inherent to the nitrogen cycle, shown in Figure 20, form the bases for nitrogen removal by biological treatment, although some removal results from nitrogen incorporation into cellular protoplasm. A major advantage of carrying out biological nitrogen removal is that all forms of nitrogen ultimately are removed from solution in the non-polluting form of nitrogen gas. The following two successive steps are involved in the biological removal of nitrogen:

1. nitrification - the bio-oxidation of
     ammonia forms to nitrate.

2. denitrification - the subsequent biolog-
     ical reduction of nitrate
     to nitrogen gas.

206

1.) NITRIFICATION. The oxidation of ammonia to nitrate is a two-step process, if carried to completion. Initially, ammonia is oxidized to nitrite by the genera of strict aerobic autotrophic bacteria *Nitrosomonas* which utilizes ammonia as its sole source of energy. The stoichiometry of this reaction is shown in Equation 9.1. The second step, the conversion of nitrite to nitrate, is accomplished by the *Nitrobacter* genera, a specific group of autotrophic bacteria using nitrite as its sole energy source. The stoichiometry of this reaction is shown in Equation 9.2. The overall nitrification reaction is represented by the summary Equation 9.3. This overall reaction illustrates the oxygen demand of ammonia nitrogen: 1.0 mg/ℓ of ammonia nitrogen (as N) requires 4.6 mg/ℓ of dissolved oxygen for complete nitrification, as indicated by a ratio of molecular weights (64 gm oxygen required per 14 gm nitrogen present).

$$2\,NH_4^+ + 3\,O_2 \rightarrow 2\,NO_2^- + 2\,H_2O + 4\,H^+ \qquad (9.1)$$

$$2\,NO_2^- + O_2 \rightarrow 2\,NO_3^- \qquad (9.2)$$

$$NH_4^+ + 2\,O_2 \rightarrow NO_3^- + 2\,H^+ + H_2O \qquad (9.3)$$

2.) DENITRIFICATION. The biological denitrification process is carried out anaerobically by heterotrophic microorganisms which utilize nitrate as a hydrogen acceptor and require an organic energy source. Denitrification must be considered as a two-step process. For example, if methanol is selected as the organic carbon source to provide energy for bacterial synthesis, the following equations represent the microbial denitrification transformations:

$$NO_3^- + \frac{1}{3}CH_3OH \rightarrow NO_2^- + \frac{1}{3}CO_2 + \frac{2}{3}H_2O \qquad (9.4)$$

$$NO_2^- + \frac{1}{2}CH_3OH \rightarrow \frac{1}{2}N_2 + \frac{1}{2}CO_2 + \frac{1}{2}H_2O + OH^- \qquad (9.5)$$

$$NO_3^- + \frac{5}{6}CH_3OH \rightarrow \frac{1}{2}N_2 + \frac{5}{6}CO_2 + \frac{7}{6}H_2O + OH^- \qquad (9.6)$$

Biological denitrification has been accomplished successfully both in stirred reactors in which microorganisms were maintained in suspension with the wastewater under anaerobic conditions and in column or packed bed adsorption systems in which microbial growth takes place on the available surface area.

EXAMPLE 2. After secondary treatment, a wastewater contains 3.0 mg/ℓ of NH_4^+. How much oxygen is required to convert the ammonium ion to nitrate ion if the wastewater flow rate is 0.01 m^3/sec?

SOLUTION :

Daily

$$\text{Oxygen Req'd} = \left(\frac{3.0 \text{ mg}/\ell \text{ NH}_4^+}{18 \text{ gm/mole}}\right)\left(\frac{2 \text{ moles oxygen}}{\text{mole NH}_4^+}\right)\left(\frac{32 \text{ gm}}{\text{mole oxygen}}\right)\left(\frac{10 \text{ } \ell}{\text{sec}}\right)$$

$$= \left(106.0 \text{ } \frac{\text{mg}}{\text{sec}}\right)\left(\frac{1 \text{ kg}}{10^6 \text{ mg}}\right)\left(\frac{86,400 \text{ sec}}{\text{day}}\right)$$

$$= 9.2 \text{ } \frac{\text{kg}}{\text{day}}$$

---

b. Chemical Removal of Nitrogen

Physical-chemical methods of removing nitrogen from wastewater are in various stages of development. The best-developed processes are ammonia stripping and breakpoint chlorination. In addition, ion-exchange methods can be utilized.

1.) Ammonia stripping [10]. Ammonium ions exist in equilibrium with ammonia and hydrogen ions in accordance with the following equation:

$$NH_4^+ \rightarrow NH_{3(gas)} + H^+ \tag{9.7}$$

With increasing pH above 7.0, the equilibrium is displaced to the right until virtually all the nitrogen will exist in the form of molecular ammonia. Figure 86 illustrates the relationship between pH and percentage of ammonia at various temperatures. Since ammonia is volatile, it can be removed by 'stripping' the water with air. General operating parameters would be to adjust the pH to the range of 10.0 to 10.5 and utilize air-liquid loadings in the range of 1500 to 2000 m^3/m^3 and aeration times in the neighborhood of 1/2 to 1 min.

Although the concept of ammonia removal by air stripping is technically and economically sound, practical applications involve some restrictions. Cold-weather operation requires special operation; blower-type operation in below freezing temperatures, for example, is not pos-

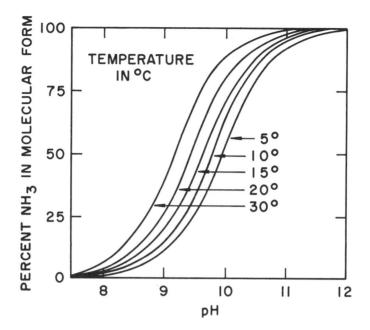

FIGURE 86. Relative Ammonia Percentages vs. pH
at Varying Temperatures.

sible, and the practice of transferring ammonia from a wastewater to the air is questionable—particularly near large bodies of water.

2.) Chlorine oxidation of ammonia [10 − 11]. It has long been recognized that chlorination during water purification treatment has resulted in oxidation of ammonia. Equation 9.8 illustrates the reaction to form hypochlorous acid from chlorine gas. In the hypochlorous acid molecule, chlorine exists in its highest oxidation state, Cl^+. It is, therefore, a very strong oxidizing agent.

$$Cl_{2(gas)} + H_2O \rightarrow HOCl + H^+ + Cl^- \qquad (9.8)$$

In drinking water, at low ammonia concentrations, hypochlorous acid reacts with ammonia to form chloramines via the following reactions:

$$NH_4^+ + HOCl \rightarrow NH_2Cl + H_2O + H^+ \qquad (9.9)$$

$$NH_2Cl + HOCl \rightarrow NHCl_2 + H_2 \qquad (9.10)$$

$$NHCl_2 + HOCl \rightarrow NCl_3 + H_2O \qquad (9.11)$$

In breakpoint chlorination, chlorine is added to process waters until a point is reached where the total dissolved residual chlorine has reached a minimum (the breakpoint) and the ammonia-nitrogen content has disappeared. Palin [12] has suggested that two reactions describe the destruction of ammonia-nitrogen:

$$NH_2Cl + NHCl_2 \rightarrow N_2 + 3\,H^+ + 3\,Cl^- \qquad (9.12)$$

$$2\,NH_2Cl + HOCl \rightarrow N_2 + 3\,H^+ + 3\,Cl^- + H_2O \qquad (9.13)$$

In practice, breakpoint chlorination of wastewater does not quite follow the simple paths represented by Equations 9.12 and 9.13. The reactions become very much more complicated. In wastewaters, not only may the ammonia-nitrogen concentration be more than an order of magnitude higher than those normally encountered in natural waters, but also numerous other dissolved and suspended species may be present in concentrations to affect significantly the basic chemical reactions between chlorine and ammonia nitrogen. A typical breakpoint chlorination curve for a secondary effluent is shown in Figure 87. The breakpoint, in this case, can be observed to occur between the 8:1 and 9:1 weight ratio of chlorine/ammonia nitrogen. Palin [12] has offered an explanation of the typical breakpoint curve. In the first ascending branch of the curve, chlorine is present in combination with ammonia as mono- and di-chloramine. The peak of the curve represents the point where all the ammonia initially present is combined to form the chloramines. In the presence of excess chlorine, the chloramines are unstable and react with free chlorine; the net result is that the amount of available chlorine is decreased. This accounts for the descending leg of the curve.

When all the chloramines have been converted to a form that will no longer react with free chlorine, the further addition causes a proportionate rise in residual chlorine. This accounts for the second ascending branch of the curve. The overall reaction of chlorine with ammonia can be expressed by the following equation:

$$2\,NH_3 + 3\,Cl_2 \rightarrow N_2 + 6\,H^+ + 6\,Cl^- \qquad (9.14)$$

From Equation 9.14 it can be seen that the final products of the reaction at the breakpoint are nitrogen gas and hydrochloric acid. Also, this is the reaction that is used normally to calculate the stoichiometric weight of chlorine needed to oxidize ammonia nitrogen.

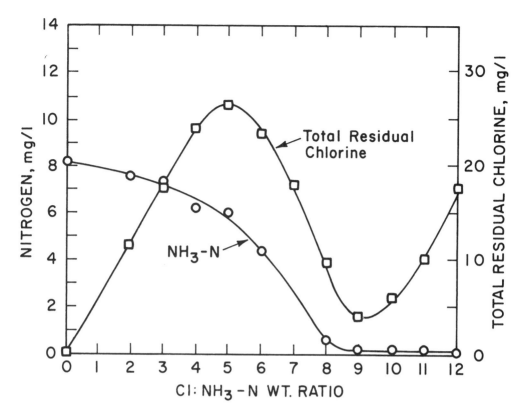

FIGURE 87. Breakpoint Chlorination of
for Secondary Effluent.

Figure 87 shows also the reduction of ammonia-nitrogen concentration with increasing chlorine concentration. Nearly all ammonia nitrogen is removed at the breakpoint. Measurements by Pressley, *et al.* [11] indicate that ammonia-nitrogen removal occurs by oxidation of the ammonia chiefly to nitrogen gas with only small amounts of nitrate-nitrogen and nitrogen trichloride being formed.

The process, basically, is selective for ammonia-nitrogen and, for the most part, will not remove the other significant nitrogen forms in water (i.e., organic, nitrite, and nitrate). Other wastewater treatment processes that would be used prior to breakpoint chlorination (e.g., activated sludge, carbon adsorption, etc.), however, successfully remove much of the organic nitrogen; nitrite- and nitrate-nitrogen concentrations are present only under highly oxidized conditions. Potential products of N_2O, O_2 (from decomposition of N_2O), NO, and NO_2 do not occur. The pH of the solution greatly influences the amount of chlorine required to reach the breakpoint for a given initial ammonia-nitrogen concentration. Pressley reported that the minimum chlorine dosage required to reach the breakpoint occurred in the range of pH 6 to 7. The pH of the solution will also greatly influence the relative concentration of each of the chloramine species and the formation of the undesirable side reactions of nitrate and nitrogen trichloride.

3.) Ammonia removal by selective ion exchange [10, 13]. The fact that ammonia exists in solution predominantly as the positively charged ammonium cation, (NH_4^+), at pH values less than about 9.5 permits it to be removed by conventional ion-exchange methods. Synthetic ion-exchange resins are generally considered to be uneconomical for the removal of ammonium ion, however, because of the selectivity of these ion-exchange resins for other cations, principally divalent ions, which predominate in wastewaters and are removed preferentially. The selectivity of synthetic ion-exchange resins follows the Hofmeister, lyotropic series $(Ca^{+2} > Mg^{+2} > K^+ > NH_4^+ > Na^+)$ which makes these exchanges inefficient in removing ammonium ion in the presence of divalent cations. Recent reports [13, 14] indicate that these problems may be minimized by using an ammonium ion-selective zeolite such as clinoptilolite. Clinoptilolite is a natural zeolite with its selectivity derived from the pore size and from the exchangeable cation sites which are selective for $K^+ > NH_4^+ > Na^+ > Ca^{+2} > Mg^{+2}$. The combination of selectivity and an exchange capacity of about 2.0 meq/gram appears to make clinoptilolite a useful exchanger for ammonium ion in wastewater. The loaded clinoptilolite can be regenerated by any basic solution at a pH value exceeding 11.0. Ammonia can be removed effectively from the regenerant by air stripping.

3. Removal of Other Constituents - Effluent Polishing

By the time wastewater has been subjected to primary, secondary, and tertiary treatment, it has been purified to a high level of water quality. In some cases, however, further consideration must be given to removing residual suspended solids, refractory organic material, and/or additional dissolved inorganic constituents. Numerous processes have been suggested for each of these purposes and many have been employed successfully in various installations. Microscreening and filtration (diatomaceous earth or granular media) processes, for example, have been successfully applied for the removal of suspended solids. In addition to the use of activated carbon for the removal of organic material, oxidative processes (e.g., ozone, molecular oxygen, catalytic oxidation with molecular oxygen, chlorine, electro-chemical treatment, etc.), and foam separation techniques have also all been used successfully.

No doubt to an increasing degree in future applications, the removal of appreciable quantities of dissolved inorganic constituents from wastewaters will be necessary. A number of different processes have been investigated for this purpose, and among those processes presently considered useful are: ion exchange, electrodialysis, reverse osmosis, and distillation. The application of each of these four processes for the treatment of wastewaters has been studied in some detail during the past years and numerous publications are available on each. Two relatively early government publications (1965 and 1968) on advanced waste treatment [15, 16] serve as good summary reports on these and other related advanced wastewater treatment processes or effluent polishing processes. The application of these special treatment processes will be presented in the next chapter in discussions on water reuse.

It is possible to produce a final water of virtually any desired quality (e.g., drinking water, deionized water, etc.) from any of the processes. The construction and operation of each of these processes, however, will remain a relatively expensive proposition. It should be noted that each of these advanced wastewater treatment processes usually requires some degree of pretreatment. Such process deterrents as fouling of resins, membranes, and/or heating elements from the presence of such impurities as suspended solids and/or dissolved organic material are common. Hence, at least primary and secondary treatment processes are required prior to the unit, and frequently carbon adsorption and filtration must also be included. The cost of this pre-treatment must also be included in the total cost figures to account somewhat for the high cost. In addition, it should also be noted that a concentrated solution, or brine, results from each process, and consideration must be given to its subsequent handling.

a. Ion Exchange

Ion exchange is a well-known method for softening or for demineralizing water. Although softening can be useful in some instances and ion exchange can be useful for ammonia removal, the most likely application for ion exchange in wastewater treatment is for demineralization. Demineralized water produced by this process is of higher quality than generally necessary, however; therefore, only part of the total stream need be treated.

Many natural materials and, more importantly, certain synthetic materials can exchange ions from an aqueous solution for ions loosely bound to the resin material itself. Cation-exchange resins, for example, can replace cations in solution with hydrogen ion from the resin. Similarly, anion-exchange resins can replace anions in solution with hydroxyl ion. A combination of these cation- and anion-exchange treatments results in a high degree of demineralization.

Since the exchange capacity of ion-exchange materials is limited, they eventually become exhausted and must be regenerated. The cation resin is regenerated with an acid; the anion resin is regenerated with a base. Important considerations in the economics of ion exchange are the type and amounts of chemicals needed for regeneration. Regeneration of the resins is an expense which cannot be avoided, and the cost of chemicals required for this purpose will represent a significant portion of the operating costs. Unless these regenerating chemicals, which themselves could create secondary pollution, are recovered and reused, this cost can be prohibitively high.

b. Electrodialysis

The principle of operation of an electrodialysis system is to impose an electric potential across a tank containing the wastewater. This will induce the dissolved ions to migrate to the oppositely charged electrode. A schematic diagram of such a system is shown in Figure 88.

This figure illustrates the alternate placement of selective membranes (i.e., cation permeable, anion permeable, respectively) between the electrodes resulting in alternate compartments becoming more concentrated with ionic species, while the intervening compartments become better (purer) in water quality due to smaller concentrations of dissolved constituents being present. A water quality approaching that of demineralized water can be achieved, but only partial demineralization is practical because electrical power requirements become excessive if the ion concentration is reduced too much.

c. Reverse Osmosis

A phenomenon known as osmosis occurs when solutions of two different concentrations are separated by a semipermeable membrane such as cellophane. With such an arrangement, water tends to pass through the membrane from the more dilute side to the more concentrated side and produce concentration equilibria on both sides of the membrane. The ideal osmotic membrane permits passage of water molecules but prevents passage of dissolved materials. The driving force that impels this flow through the membrane is related to the difference in concentration between the two solutions, and an osmotic pressure difference between the two compartments results. The principle is illustrated in Figure 89A. If the liquid on the more concentrated side is allowed to

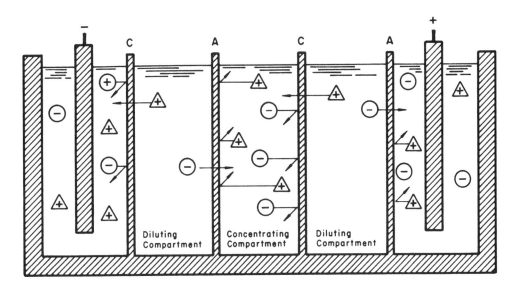

FIGURE 88. Schematic Representation of Electrodialysis;
A—Anion-Permeable Membrane; C—Cation-permeable Membrane;
Θ—Anion; Δ—Cation.

rise in a standpipe as water passes through the membrane, as shown in Figure 89B, the hydrostatic pressure on the right side of the membrane gradually increases until finally it equals the osmotic pressure. At this point, flow through the membrane ceases. If, then, the pressure on the more concentrated (right) side of the membrane is increased, as shown in Figure 89C, the flow of water through the membrane reverses; that is, water moves from the more concentrated compartment to the less concentrated compartment. This method of wastewater treatment is known as reverse osmosis.

Films of several types of material have the property of preventing the passage of minerals while allowing the passage of water in which the minerals were dissolved. This process, however, is relatively expensive and is usually applied for the production of specialty waters.

d. Distillation

Distillation has long been known as a process for producing 'pure' water. It is capable of removing both organic and inorganic dissolved contaminants with relatively high efficiency; suspended solids are eliminated and even microorganisms are destroyed by the boiling temperatures employed. Volatile impurities in the feed pass over into the distillate, but several 'polishing' techniques are available to eliminate this problem.

215

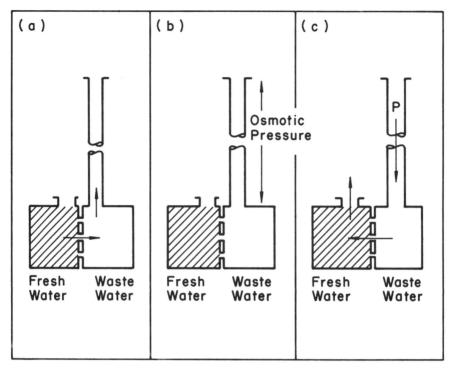

FIGURE 89. Schematic Principles of Osmosis;
(a) Normal Osmosis, (b) Equilibrium, and (c) Reverse Osmosis.

Compared with the 'organic only' or 'inorganic only' removal processes, distillation remains a relatively expensive process, even though much recent work in the desalinization field has been aimed at decreasing costs. The prospects of lower cost nuclear energy or combination water renovation/power plants might improve the economics significantly in the future.

Figure 90 illustrates a multiple-effect evaporator, commonly used to repeatedly capture the latent heat of vaporization.

In this system, each effect is maintained at a slightly lower pressure and, hence, a lower temperature than the preceding one. This allows the steam produced in one effect to be used as the heating medium for the next. The result is that for each kilo of steam supplied to the first effect, there is produced a number of kilos of product water approximately equal to the number of effects. Other types of distillation equipment available include multi-stage flash evaporators and vapor-compression distillation.

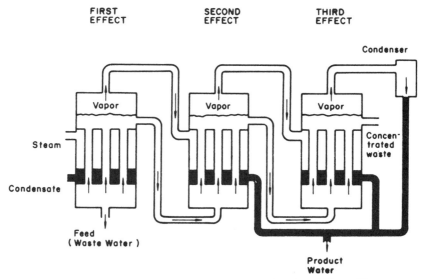

FIGURE 90. Schematic Diagram of Multiple-Effect Evaporator.

REFERENCES

[1] W. W. Weber and E.H. Smith, *Environmental Science and Technology*, **20**, No. 10, 970 (1986).

[2] W. J. Weber, *et al.,J. Water Pollution Control Federation*, **42**, 83 (1970).

[3] E. F. Barth, *et al.*, *J. Water Pollution Control Federation*, **40**, 2040 (1968).

[4] D. Vacker, *et al.,J. Water Pollution Control Federation*, **39**, 750 (1970).

[5] J. L. Witheron, *Proceedings of 24th Industrial Waste Conference*, Purdue University, 1969.

[6] W. F. Milbury, *et al.,J. Water Pollution Control Federation*, **43**, 1890 (1971).

[7] R. D. Borgman, *et al.*, Paper Presented at 5th International Water Pollution Research Conference, San Francisco, CA, 1971.

[8] J. B. Carberry and M. W. Tenney, *J. Water Pollution Control Federation*, **45**, 2444 (1973).

[9] L.G. Sillen and A.E. Martell, *Stability Constants of Metal-Ion Complexes*, Special Publication (No. 17) Chemical Society of London, 1964.

[10] J. M. Cohen, in *Nutrients in Natural Waters*, H. E. Allen and J.

R. Kramer, eds., Wiley Interscience, New York, 1972.

[11] T. A. Pressley, *et al.*, *Environ. Science & Technolology*, **6**, 622 (1972).

[12] A. T. Palin, *Water & Wastes Eng.*, **54**, 151, 189, and 248 (1950).

[13] S. R. Linne and M. J. Semmens, *Proceed. 39th Purdue Industrial Waste Conference*, Ann Arbor Sci. Publ. Woburn MA, **39**, 757 (1985).

[14] B.W. Mercer, *et al.*, *J. Water Pollution Control Federation*, **42**, R95, (1970).

[15] Summary Report - The Advanced Waste Treatment Program (PHS Publication No. 999-WP-24), April, 1965.

[16] Summary Report - Advanced Waste Treatment (WPCA Publication No. WP-20-AWTR-19), 1968.

HOMEWORK

1. Calculate the required daily moles of methanol, CH_3OH, for the wastewater in Example 2 if the denitrification process is to be carried out. HINT: Use reactions in Equations 9.3 and 9.6.

2. Calculate the daily molar chlorine requirement for the wastewater in Example 2 if the nitrogen removal is due to chlorine oxidation. HINT: Use reaction in Equation 9.14.

3. Calculate the daily required grams of artificial zeolite for nitrogen removal from the wastewater in Example 2 if the ion exchange capacity is 3.0 mili-moles per gram of resin.

Chapter 10

WATER REUSE AND
DRINKING WATER TREATMENT

A. Historical Considerations

The term "re-use" has become more commonplace in the last decade than previously. Several factors have contributed to the popularization of this term. The environmental movement of the 1960's led to the passage and implementation of several federal laws such as the Clean Water Act of 1972 and the Safe Drinking Water Act of 1979. These laws mandated the federal Environmental Protection Agency to financially assist states and their municipalities in cleaning up wastewater discharges to the environment and in providing reliable, safe drinking water, respectively.

The Clean Water Act of 1972 required the EPA to establish a set of discharge permits for every discharge to interstate waterways. These discharge permits were negotiated individually by the federal EPA based on federal economic technological standards. This system of permits is referred to as the NPDES system (National Pollution Discharge Elimination System). Municipalities were awarded federal money to design and construct new secondary wastewater treatment plants mandated by the law. Industries were given the following options: 1.) discharging treatable wastes to the local wastewater treatment plant and paying user charges based on the industrial flow and pollutant concentration discharged, or 2.) building and operating their own on-site treatment plants with no financial assistance from the federal government. Before this law was passed in 1972, individual states were reluctant to pass strict pollution discharge laws for fear that industry would then leave the state for an area where more lenient laws prevailed. Often the wealthiest, most industrially-developed states had the poorest water quality. For example, it was reputed that the Cayahoga River south of Cleveland, Ohio caught on fire. In addition, the individual states had no jurisdiction over interstate waterways. It was also claimed that vast stretches of the great Mississippi River were dead due to uncontrolled discharges of municipal and industrial waste. And, consequently, such downstream states, such as Louisiana, argued that upstream states had reduced the river use to transportation only by the time it reached their jurisdiction. (See Table 6)

The philosophy of the Clean Water Act of 1972 was new and revolutionary at the time. The idea involved a deal – the states would accept the imposition of uniform federal regulations in exchange for financial funding to build wastewater treatment plants. Since this law was passed, the individual state regulatory agencies have gradually taken over the implementation of this law and managed the monitoring and permit renewal programs which the law requires. Some states have even proposed more strict regulations than the federal EPA has established. Some states also have imposed similar regulations on discharges to their intrastate waterways. Industry, with few exceptions, has complied with this very complex law and its periodic Congressional updates.

The Safe Drinking Water Act of 1979 was less revolutionary than the Clean Water Act of 1972, but no less vast in scope. This law required every municipality or private drinking water supply company which distributed drinking water to its citizens or consumers of 15 households or greater to comply with the U.S. Public Health Service Drinking Water Standards. (See Table 4.) This law superficially appears to be very straightforward, but it has led to the realization that many communities do not have satisfactory drinking water. Some communities experience variations in water quality intake to the drinking water treatment plant or intermittent lapses in drinking water treatment processes within the plant [1]. In addition, many community officials have been dissatisfied that the Drinking Water Standards have not been updated and approved since 1962. Consequently, recent epidemiological data are not encompassed by the Drinking Water Standards, and upgrading existing drinking water treatment plants to comply with the 1962 Standards may become only a short-term solution. Furthermore, the Safe Drinking Water Act has not been funded at the same financial level as the Clean Water Act. The expense of upgrading drinking water plants must be borne mostly by the local authority, whereas the upgrading of wastewater treatment plants has been financed largely by the federal government.

The final historical factor to consider is the realization that many drinking water sources are contaminated with toxic and/or carcinogenic pollutants. Such pollutants can enter the surface and groundwaters from various sources. These sources have become classified as point sources, such as an outfall from a wastewater treatment plant to a receiving stream, or from a leaking underground storage tank at an industrial site which slowly discharges an uncontrolled plume that eventually trickles down to an underground aquifer. Most toxic organic chemicals which might enter the potential drinking water sources have very limited solubility in water. Often it is difficult to measure the concentration or even detect the presence of these chemicals at the very low concentrations which are actually toxic or carcinogenic; for many of these chemicals these levels have not been determined. All of these contingencies are reflected in Table 4 in that the noted additions to the U.S. Public Health 1962 Drinking Water Standards have never been finally approved. The required approval process is painstakingly slow and various responsible public authorities cannot agree on either the chemical contents of the standards or the permitted concentration levels of each. Since portions of the Drinking Water Standards are not finally approved, the standards, as shown, are referred to as Interim Drinking Water Standards. Most state regulatory agencies require or at least encourage their municipalities to comply with these interim standards for the treated drinking water which they distribute to their consumers. Many consumers, however, are dissatisfied, and purchase bottled water. Some homeowners install carbon adsorption capability to remove potentially toxic or carcinogenic dissolved organics, or they may install ion exchange units for water softening purposes. Citizens appear to be more conscious of links between environmental quality and health effects. Municipal officials and engineers have become more sophisticated, too, and aware of the health limitations of conventional drinking water treatment. Many municipalities have been required to increase user charges for both drinking water treatment and wastewater treatment in order to comply with the Safe Drinking Water Act and the Clean Water Act. Some officials have considered the possibility of direct water reuse as the regulations on wastewater discharge quality grow more and more stringent, while the quality of available drinking water resources continues to deteriorate or become less reliable.

B. Direct Reuse Considerations

Many surface water resources are reused indirectly over and over. Indirect reuse occurs when a municipality withdraws water from a river or stream, treats the water to provide drinking water for its community, collects the used water and treats this wastewater before discharging it downstream from the community location. The next downstream municipality will repeat this sequence of withdrawal, treatment, use, treatment and discharge. Since wastewater treatment plants cannot remove 100% of pollutants and are, in fact, permitted to discharge a small concentration of both BOD and suspended solids, then the sequential water indirect reuse causes an increasing deterioration of water quality downstream. Reference has already been made to the case of the Mississippi River.

The question for municipal officials to consider is: Which drinking water source is cheaper, wastewater treated to a level of quality to achieve drinking water standards or the available drinking water source to be treated to achieve a quality which complies with drinking water standards? In addition, Table 6 presents water priority uses which indicate that wastewater could be treated to a level of quality lower than that required for drinking water purposes and still be reusable directly for manufacturing or agricultural purposes.

It is useful to look at the different classes of users and their volumetric demand. In the U.S., the largest demand is for agricultural purposes. This demand is often seasonal, and the period of greatest demand is often during drought periods when the supply is meager. Water demand for agriculture is greater in the western part of the U.S. than in the east.

The second largest water user is industry. Industrial water requirements are primarily for cooling water used in the power generating industry and, secondarily, for process stream water used in manufacturing industries. In some cases, the required water quality for these two industrial users is too high to permit reuse of treated wastewater. In other cases, treated wastewater is of higher quality than any other convenient water source available to industry. An example of such serendipitous reuse is the 30 MGD (millions of gallons per day) or 1.3 m^3/sec of secondary effluent which the Sparrows Point Works of Bethlehem Steel purchases from the Baltimore, MD, Back River Wastewater Treatment Plant. This effluent is of such high quality that it is used by the steel plant for finishing water in the steel rolling process. It is then retreated on the industrial site and discharged into the outer Baltimore Harbor.

222

Industrial water demand is greatest in the eastern U.S., rather than in the west. Industrial sites are often nearby metropolitan areas so that potential industrial reusers could be conveniently located for receiving wastewater treatment plant effluents. Both the power generating industry and manufacturing industries have determined, however, that once they treat their used water, it is more economical to recycle this water onsite, rather than discharge it to a receiving water. This result of the Clean Water Act industrial requirements has led, therefore, to an unanticipated reduction in industrial water demand.

The third category of potential reusers is through groundwater recharge. The recharge may be accomplished by spraying the wastewater treatment plant effluent over permeable soil and allowing the liquid to percolate through the soil. The liquid interacts with the soil particles, with dissolved constituents undergoing adsorption or ion exchange processes with the soil particles. Any carbon-containing pollutant may induce aerobic or anaerobic metabolism in microbes occluded in the soil strata or attached to the soil particles themselves. These processes continue as the liquid continues to percolate downward through the soil column. Eventually the liquid plume enters an underground aquifer where the soil stratum is saturated with water.

Until recently, these underground aquifers were considered as a purer source of drinking water supply than surface waters. Now these drinking water sources have been tested and often found to contain toxic and carcinogenic chemicals present from leaking storage tanks or improperly sealed waste sites. It is important, therefore, to realize that the introduction of recharged wastewater treatment plant effluents must not threaten the water quality and potential use of the receiving underground aquifer. In addition, all underground aquifers eventually flow toward and are discharged to surface waters. Any contamination of an underground aquifer, therefore, will affect the water quality of its receiving surface water. That these water sources are all linked can be demonstrated by measuring a persistent chemical such as DDT in representative locations of different reservoirs presented in Table 2. Previous to this realization of linked water resources, often industries were permitted to discharge wastes to ponds containing no impenetrable barrier which is now required. The liquid could percolate into the surrounding soil, eventually penetrating an underground aquifer and then subsequently contaminate its receiving surface water. In some cases involving particularly objectionable wastes, some industries practiced deep-well injection to confined underground strata. This disposal technique was at the time considered to be a permanent disposal of the waste and its threat to human health and the environment was considered to be minimal. Toxic waste disposal in sealed steel drums was also considered an acceptable ultimate disposal technique in the past.

With the recent emphasis on protecting the quality of both our underground and surface water resources, the opportunity to develop direct water recycle for drinking water purposes is shrouded with apprehension. All circumstances must combine to favor direct recycle; i.e., drinking water sources must be scarce or polluted to levels which cannot be treated to comply with the U.S. Public Health Service Drinking Water Standards. Recent reports to the World Health Organization describe a project in Saudi Arabia to reuse treated effluent, mostly for irrigation purposes. As much as 10% of this flow, however, can be treated further in order to reuse directly as drinking water[2].

Drinking water sources are presently very scarce in the southwestern U.S. In some areas of this region, shortfalls already exist and are expected to grow worse. Direct reuse for drinking water purposes, however, is not expected to increase significantly in the U.S. during the 1990's. Instead, recycling and reuse is expected to increase within industry, both the power-generating industry and the manufacturing industry. Any decrease in municipal demand resulting from the reduced intake by industry will provide greater allocations from existing water resources for drinking water purposes. In addition, it is anticipated that increased wastewater treatment plant effluent reuse will be developed for industry and agriculture. A modified figure from Montgomery [3] is presented in Figure 91 to illustrate the potential reuse hierarchy related to the degree of wastewater treatment imposed.

Figure 91 indicates that much more efficient use of our water resources can be developed if the effluent reuser is conveniently located and if proper planning and implementation can occur. As previously discussed in Section A, the Clean Water Act requires that primary and secondary treatment be carried out on all municipal wastewaters. The top two types of direct reuse could be carried out, then, on effluents from wastewater treatment plants which presently comply with their discharge permits issued under this law. The third level of direct reuse could be developed for effluents from wastewater treatment plants which carry out either imposed or voluntary tertiary treatment. Perhaps the greatest volume of effluent direct reuse could occur after the tertiary treatment stage.

Processes cited for the effluent polishing stage are very expensive and are not carried out routinely at wastewater treatment plants. Normally then, the tertiary-treated effluent intended for direct reuse would

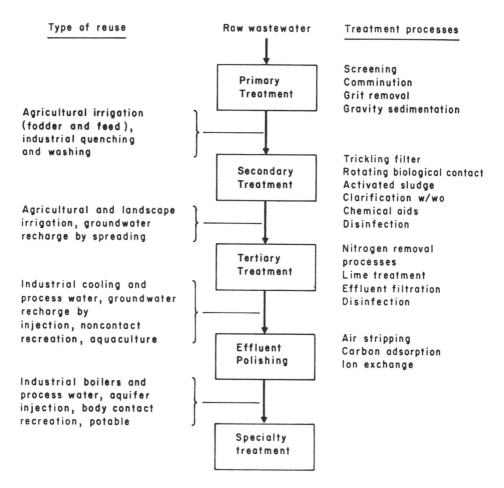

Type of reuse   Raw wastewater   Treatment processes

FIGURE 91. Hiarchical Relationship of Treatment Stages
and Reuse Strategies.

be transported off-site from the wastewater treatment plant to an industrial site or to a drinking water treatment plant site. Such processes are presented in detail in Chapter 9 and would most likely be carried out on site as an industrial pre-treatment procedure or implemented in preparation for drinking water treatment at a drinking water plant.

C. Drinking Water Treatment

Drinking water treatment, in contrast to wastewater treatment, is quite standardized; the variation in treatment plant schemes arises from the different types of sources, 1.) surface water, or 2.) underground aquifer. If the re-use of reclaimed wastewater is to be considered as a drinking

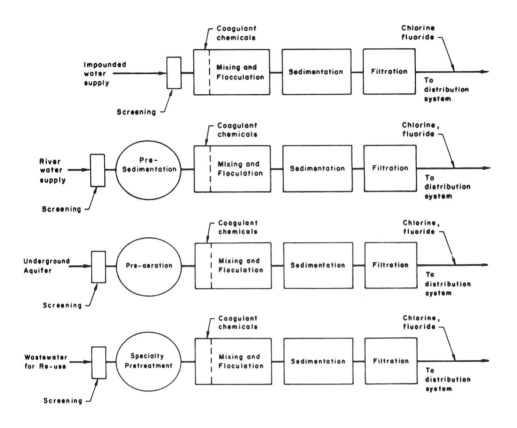

FIGURE 92. Drinking Water Treatment for various Types of Resources.

water source, then this third class of sources must be included. A modification of a flow diagram from Viessman and Welty [4] is shown in Figure 92. From this figure one can note that the increase in drinking water treatment complexity progresses from that level required to treat groundwater sources, through the level of complexity required for treating surface water sources, to that for treating reclaimed wastewater.

Two additional differences between drinking water treatment and wastewater treatment are as follows:

1. Drinking water treatment is designed and operated to provide the most careful, risk-free prevention of water-borne disease transmission. Wastewater treatment is intended only to economically minimize the impact of BOD and suspended solids discharge to the environment.

226

2. The quality and flow of raw wastewater into a wastewater treatment plant varies considerably during any 24 hour period and during various weather events. The drinking water treatment plant, on the other hand, can take in water at a steady flow rate continuously from a source and with predictable water quality. The quality of groundwater drinking water sources hardly varies. Surface waters used for drinking water sources will vary predictably with storm events when suspended solids concentrations will be high.

3. If reclaimed wastewaters are to be used as a drinking water source, this water quality will vary unpredictably at the intake of a drinking water treatment plant. The processes implemented as effluent polishing or drinking water pretreatment processes will serve to minimize the water quality variations undergoing conventional drinking water treatment. If the drinking water treatment processes are properly designed and operated, then the risk of disease transmission from using reclaimed wastewater may be reduced to a level comparable to the risk associated with groundwater sources and surface water sources.

Examination of the unit processes illustrated in Figure 92 reveals that many of the processes have been presented already in Chapters 6 and 7 for wastewater treatment. It should be emphasized that these processes are the same processes used for primary and secondary wastewater treatment, such as sedimentation and clarification. The important additional process shown in Figure 92, the filtration process, is the principal process in drinking water treatment.

The remainder of this chapter, then, will briefly discuss the applications of unit processes in drinking water for ground water sources, surface water sources and reclaimed wastewater sources. Following this section, the fundamental theory of the filtration process will be presented. And, lastly, some design of these facilities can be illustrated.

1. Treatment of groundwater for drinking water:

Aeration - groundwaters usually contain a low DO concentration and pH. Under these conditions most metals exist in a reduced state which is more soluble than a highly oxidized state. Aeration, using mechanical or diffused aeration systems will provide oxygen to oxidize metals such as iron and manganese which are undesirable in water supplies. The reaction for the oxidation of iron is as follows:

$$2\,Fe^{+2} + 6\,H^+ + 3\,O_2 \longrightarrow 2\,Fe(OH)_3 \qquad (10.1)$$

It should be noted that in the presence of bicarbonate alkalinity, ferric carbonate may form and be precipitated. Furthermore, reduced ferrous ion can form complexes with organics and these cannot be oxidized and easily removed. The ferric hydroxide, formed in Equation 10.1 and ferric carbonate are very insoluble, however. These forms precipitate out of solution and can be removed in the subsequent coagulation, sedimentation and filtration processes.

The oxidation of manganese is more complex than oxidation of iron. Manganese must be oxidized from a (+2) state to a (+4) oxidation state. The loss of two electrons from the manganous ion takes a longer time period than for the transfer of one electron in the case of iron. Manganous oxidation forms an intermediate oxidation state then which may also be insoluble to some extent. In addition, manganous ion in the (+2) state can, like iron, also form complexes with organic constituents dissolved in the water and never be oxidized and removed.

The principal idea, however, for the removal of these undesirable ions is to oxidize the soluble forms to higher oxidation states which are insoluble. The oxidation and subsequent removal of the resulting precipitates in a drinking water treatment plant prevents the formation of these precipitates and their discoloration effects during the household laundry agitation process which promotes aeration and oxidation of these ions to cause stains. The aeration of groundwater supplies in a drinking water plant likewise prevents the formation of these precipitates during various industrial processes. The formation of these iron and manganese precipitates might impose undue wear on industrial machines and pumps used during manufacturing processes. The aeration process carried out in a drinking water treatment plant therefore eliminates complaints due to undesirable effects of precipitates formed during household and industrial usage.

Screening - Screens used in drinking water treatment plants are usually fine mesh screens, rather than coarse bar screens installed in wastewater treatment plants. These screens in a drinking water treatment plant must be designed so that the flow velocity is great enough to prevent sedimentation of settleable particles in the channel. In order to prevent reduction of the flow velocity, the maintenance of the screens must be vigilant to prevent clogging of the openings and the resulting increase of pressure differential between the upstream and downstream side of the screen. Usually the intake for a drinking water treatment plant contains smaller sized solid particles than the sizes prevalent in a wastewater treatment plant influent.

Mixing and Coagulation - particles in a drinking water intake are, in

228

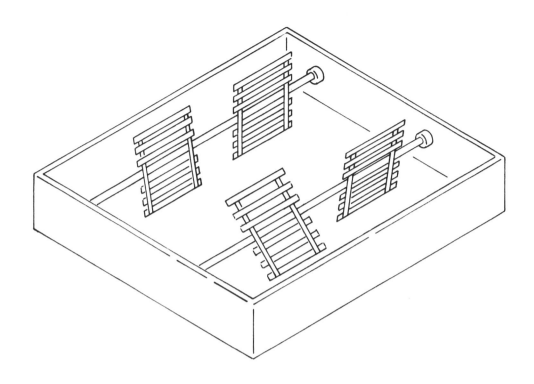

FIGURE 93. Paddle Flocculator.

fact, very small - in the colloidal range - and often require the addition of chemicals to induce coagulation. The chemical addition must take place in a rapid mixer to induce rapid distribution of the coagulant chemical. The coagulation process itself must take place under slow mixing in tanks which induce a velocity gradient in the slowly mixed water. The dual-purpose tank is usually divided into 2 sections, one complete mix tank for addition of the optimum chemical dose as discussed in Chapter 3 and one long plug flow section containing slow-moving horizontal shaft paddle mixers. An illustration of such paddle mixers for coagulation is shown in Figure 93 [3].

Usually, empirical specifications are utilized to carry out the tank design for the coagulation process. Recommended detention time for this process is 20-30 minutes, and the tank volume, V, is determined from Equation 6.12, $V = Q\Theta$. The plug flow flocculator must possess a long, narrow geometry and shallow depth. Since slow mixing must be carried out in this tank to provide a velocity gradient, the degree of ideal plug flow through the tank is diminished. It has been shown that often the coagulation performance is more effective if the total process volume required is divided into a series of separate tanks in order to promote plug flow [5]. The adverse effects of slow mixing on the reduction of plug flow hydraulics can also be reduced by inserting baffles perpendicular to the direction of flow at a few locations down the length of the tank.

Once the optimum chemical dose has been determined by jar tests, as described in Chapter 3, and once the tank volume and configuration has been decided, the operation of coagulating slow mixers can be estimated by using the following empirical equations:

$$P/V = \mu G^2 \tag{10.2}$$

where P = power dissipated (or required as input),
in units of joules/sec
V = tank volume, in units of m^3
μ = absolute viscosity, in units of kg/m-sec
G = velocity gradient, in units of sec^{-1}

Experience has indicated that the velocity gradient undergoes an optimum mixing velocity; for if the mixing is too slow, the chemical and particles to be coagulated do not contact each other with sufficient frequency; whereas, too great a velocity provides such a brief contact time that chemical reaction cannot occur. For various types of particles to be coagulated by various chemical coagulants, tables of optimum velocity gradients are available. For horizontal shaft mixers with inorganic particles typical of those found in drinking water sources and inorganic chemical coagulants such as lime or alum, the optimum range for G = 20-50 sec^{-1}. The absolute viscosity for water at 20 °C is μ = 10^{-3} kg/m-sec. If the tank volume, V, has been calculated from Equation 6.12, then Equation 10.2 can be used to determine the power required for optimum coagulation.

The following empirical equation can be used to determine the mixer paddle area and velocity relationship:

$$G = \sqrt{\frac{C_D A v^3}{2\nu V}} \tag{10.3}$$

where C_D = coefficient of drag, dimensionless
A = paddle area, in units of m^2
v = velocity of paddles, m/time
ν = kinematic viscosity, m^2/sec
V = tank volume

If C_D = 1.8 and ν = 10^{-6} m^2/sec, the tank volume and velocity gradient are known, then the relationship of paddle area and velocity can be determined.

EXAMPLE 1. Calculate the power required to operate a paddle flocculator if $G = 50$ sec^{-1}, $\mu = 10^{-3}$ kg/m - sec, and the volume of the flocculation tank is 30 m^3.

SOLUTION: Using Equation 10.2,

$$P = \mu G^2 V$$

$$= (10^{-3}\ \frac{\text{kg}}{\text{m} - \text{sec}})(50\ \text{sec}^{-1})^2(30\ \text{m}^3)$$

$$= 75\ \frac{\text{kg} - \text{m}^3}{\text{m}^4 - \text{sec}^3}$$

$$= 75\ \frac{\text{kg} - \text{m}^2}{\text{sec}^3}$$

$$= 75\ \text{Joule/sec}$$

EXAMPLE 2. Determine the paddle velocity required for the flocculation tank described in Example 1 if $C_D = 1.8$ and paddle area $= 20$ m^2. The tank volume $= 30$ m^3, the velocity gradient $= 50$ sec^{-1}, and $\mu = 10^{-3}$ kg/m $-$ sec ($\nu = 10^{-6}$ m^2/sec).

SOLUTION: Using Equation 10.3,

$$\sqrt{v^3} = \frac{G\sqrt{2\nu V}}{\sqrt{C_D A}}$$

$$= \frac{(50\ \text{sec}^{-1})(\sqrt{2(10^{-6}\ \frac{\text{m}^2}{\text{sec}})(30\ \text{m}^3)}}{\sqrt{1.8(20\ \text{m}^2)}}$$

$$= \frac{\sqrt{60 \times 10^{-6}\ \frac{\text{m}^3 - \text{m}^2}{\text{sec}}}(50\ \text{sec}^{-1})}{\sqrt{36\ \text{m}^2}}$$

$$= \sqrt{1.67 \times 10^{-6}\ \frac{\text{m}^3}{\text{sec}}}(50\ \text{sec}^{-1})$$

$$v^3 = (1.67 \times 10^{-6}\ \frac{\text{m}^3}{\text{sec}})(50\ \text{sec}^{-1})^2$$

$$= 4.167 \times 10^{-3}\ \frac{\text{m}^3}{\text{sec}^3}$$

$$v = 0.16\ \text{m/sec}$$

Sedimentation - the sedimentation process, the next process shown in Figure 92 for the drinking treatment train, is similar to the sedimentation process described previously in Chapter 7. In this case of drinking water treatment, the particles are flocculating or coagulating due to chemical addition as they descend, and we say that Type II sedimentation is taking place. The approach to Type II sedimentation tank design has been presented in Chapter 7, Example 5.

Filtration - the deep bed filtration process is the principal purification process for the drinking water treatment plant. This process removes suspended particles which have not settled out during the preceding sedimentation process. The filtration process takes place in a tank containing approximately 0.6 meter depth of sand and/or other solid medium. The unfiltered water is introduced to the top of the sand bed and percolates through the sand medium due to gravity. The filtered water is collected in the underdrains which line the tank and is sent on for further treatment such as chlorination for disinfection purposes.

The particles in suspension are removed from the liquid streamlines as they flow through the interstitial spaces between the sand particles in the filter bed. As the particles are removed from the liquid phase and captured in the sand bed, the porosity of the sand bed decreases. As the bed porosity decreases, the flow velocity through the interstitial pore spaces increases, but the height of standing, unfiltered water above the sand bed increases as well. As the height of water above the sand bed increases, the hydrostatic head increases and the resulting head loss build up becomes so great that the filtration process becomes very inefficient and must be halted. As the liquid flow velocity increases with increasing filter clogging, the interstitial shear forces increase and particles attached to the sand bed grains can be sheared off and re-entrained in the liquid flow. It is possible, therefore, that as the filter bed is becoming nearly clogged, particles can escape the filter bed and be collected in the underdrain system with the filtered water. A useful filter design and operation would combine the effects of maximum allowable head loss build up time with the filtration time required for particle breakthrough from the filter.

Once the maximum head loss build up or particle breakthrough occur, the filtration run through the filter bed must be terminated. The filter bed must then be cleansed by a backwash process during which a portion of the filtered water is vigorously pumped upward to provide turbulent flow through the sand bed and scour the deposited particles off the sand surfaces. Backwash water is then collected separately from the filtered water and this concentrated suspension is thickened and treated as a sludge with sludge collected in the sedimentation tank.

Design and operation of the filtration process evolved as part theoretical for microparameter considerations and part empirical for macro-

Table 12
A. Typical Design Values for Filter Bed Parameters

PARAMETER	SYMBOL	VALUE
diameter medium particle	d_m	1 mm
Length or depth filter	L	0.65 m
Surface Application Rate	Q/A	10 m/hr
Initial particle concentration	C_o	30 mg/ℓ
Deposit density	ρ_p	1-2 kg/m^3
Bed porosity	ϵ	0.4

B. Effect of Parameter Increase on

PARAMETER	t_{hl}	t_{bt}
d_m	Inc	Dec
L	Dec	Inc
Q/A	Dec	Dec
C_o	Dec	Dec
ρ_p	Dec	Dec
ϵ	Inc	Dec

parameter considerations. The process is very complicated because the deep bed filter never reaches a steady-state operation. Every position in the depth of the filter changes during the length of the filter run, t_f. In addition, with each run and backwash cleansing, the initial condition of the filter may differ during a succession of filter runs. Filtration efficiency may change, then, during the life of the filter. Theoretical filter bed design and operation is impossible to determine. Any semi-theoretical design approach must be calibrated by pilot plant operation of a model filter designed from theoretical considerations. Table 12 [3] provides a first order approximation of values for various filter bed parameters and the effects of increasing or decreasing these values.

All of these parameters would possess an optimum value for a given suspension to be filtered. The filter medium diameter of approximately 1 mm is large enough to provide a bed porosity which permits penetration of the suspended particles into the filter depth and does not just strain out the particles on top of the filter bed. A bed depth of less than 1 meter provides a tortuous path length sufficient to intercept suspended solids in the laminar flow streams and remove them from suspension. This volume of filter medium can be expanded efficiently during the upflow backwash process to clean the accumulated sludge from the filter bed. Any deeper filter bed would increase the probability of capture but increase the difficulty of backwash. The application rate for the introduction of suspended solids to a filter bed has been maintained at 5 m/hr (2 gpm/ft²) ever since filtration has been carried out on municipal water supplies. Recently, however, modifications of application rates have indicated that increasing the application rate utilizes the bed depth efficiency if the bed porosity is sufficient. The resulting head loss build up rate during a filter run is diminished.

In order to optimize all of these design and operation variables, pilot plant tests can be run to test the performance of a filter bed design and operation with the particular suspension to be filtered. In order to optimize the design and operation of the filter bed, it would be very efficient to synchronize the two effects of head loss build up and suspended particle breakthrough at the end of every filter run. If the filter run time to produce the maximum allowable head loss is t_{hl}, and the filter run time to produce particle breakthrough is t_{bt}, then the efficient synchronized design and operation would permit $t_f = t_{hl} = t_{bt}$, where t_f is the length of filter run. The headloss build up can be expressed by the Karman-Kozeny Equation, as follows:

$$\frac{hl}{L} = \frac{180\mu v_{liq}}{\rho_{liq}gd_{\mathrm{m}}^2}\frac{(1 - \epsilon_o)^2}{\epsilon_o^3} \tag{10.4}$$

where μ = absolute viscosity
v_{liq} = liquid velocity
ρ_{liq} = density of liquid
g = acceleration due to gravity
ϵ_o = initial bed porosity
L = depth of filter bed
d_{m} = diameter of filter medium particles

One can observe from Equation 10.4 that head loss increases with increasing liquid flow velocity and decreases with increasing diameter of the particles in the filter bed serving as filtration medium.

The equation governing the suspended particle removal efficiency was developed by Iwasaki [6], as follows:

$$\left(\frac{\partial C}{\partial L}\right)_t = -\lambda C \tag{10.5}$$

where C = concentration of remaining suspended solids
L = filter depth
λ = filter coefficient

This equation is in the form of a familiar first order rate expression. The expression must, however, contain a partial differential, because the concentration of suspended solids varies both with depth of filter

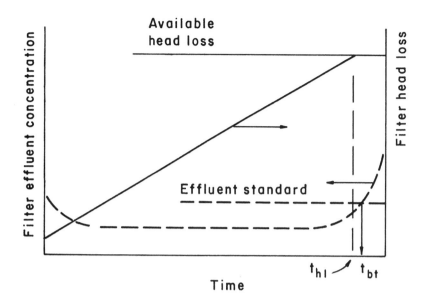

FIGURE 94. Optimization of Filter Run Time.

bed and time of filter run. This particular form of the equation is expressed as a function of filter depth at constant time, t. The partial differential equation can be integrated between any two positions in the filter bed, i and $i + 1$, increasing with depth, to give an expression for the concentration of suspended solids.

$$\ln C_i/C_{i+1} = \lambda(L_{i+1} - L_i)$$

where C_i and C_{i+1} = concentration of suspended solids
at i and $i + 1$ depths in the filter bed.

Equations 10.4 and 10.5 can be coupled to determine the optimum combination of application velocity, diameter of particle in the filter bed medium, and the corresponding bed porosity. The filter bed depth can be analyzed therefore, to determine the optimum combination of design and operation parameters which synchronize the maximum allowable head loss buildup with the onset of particle breakthrough from the filter. A computer homework problem will be assigned to apply this type of exercise illustrated in Figure 94. Emphasis should be placed on the fact that the constants for these equations must be determined empirically for each suspension.

Disinfection - In Figure 92 the last step shown in the treatment train

235

is shown to be disinfection. This process must be carried out prior to distribution to the municipality in order to provide a chlorine residual in the transmission system of 1 mg/ℓ. The optimum dose determination for this process was described in Chapter 9 using a chlorine breakpoint curve. The optimum dose can be injected into the water in the distribution system itself or in a separate, long narrow plug flow tank. The tank is designed according to Equation 6.12 in order to provide the necessary contact time, usually one half hour for a given volumetric flow rate, Q.

2. Treatment of surface water for drinking water:

Pre-sedimentation - Surface water is treated for drinking water purposes much the same as groundwater, except that in place of the aeration for ground water supplies, a pre-sedimentation process is carried out on most surface waters. Figure 92 illustrates this difference between the two treatment trains. The pre-sedimentation process is carried out especially for water taken from moving surface water sources which are subject to sediment increases during storm events. Impounded sources from reservoirs rarely require a pre-sedimentation step.

The pre-sedimentation process for surface waters in a drinking water treatment plant is exactly the same as s primary sedimentation process in a wastewater treatment plant. The suspended particles in a drinking water source, however, are smaller than those entering a wastewater treatment plant. The pre-sedimentation process in a drinking water treatment plant, therefore removes a smaller percentage of total suspended solids than the primary process in a wastewater treatment plant. The solids removed in the pre-sedimentation process remove the settleable solids in a Type I sedimentation process. The tank is designed using the same theoretical considerations and semi-empirical approaches as described in Chapter 6. In the drinking water treatment plant, however, the principal sedimentation process takes place following coagulation and utilizes Type II principals for coagulating particles.

The processes subsequent to pre-sedimentation in the treatment of surface water are the same as for the treatment of groundwater.

3. Treatment of wastewater treatment plant effluent for drinking water:

Earlier in this chapter, the use of wastewater treatment plant effluent for drinking water purposes was described as a category of water reuse. The effluent polishing processes presented in Chapter 9 can be carried out at the wastewater treatment plant prior to discharge of the effluent to the drinking water treatment plant or these processes may be carried out at the drinking water treatment plant as a pre-treatment process.

In Figure 91, these polishing processes are illustrated as a pre-treatment process. It must be emphasized that these pre-treatment processes are intended to purify the water as clean as possible prior to the drinking water treatment itself. For example, any residual suspended solids will impair the disinfection process. Such impairment may also occur in drinking water treatment plants using surface or groundwaters for the drinking water treatment plant intake. When wastewater treatment plant effluent is used as the drinking water source, however, extreme care must be taken to eliminate any potential pathogens in the form of microbial suspended solids such as bacteria, cysts, yeasts, and viruses. Once the effluents from the drinking water treatment plant have been proven to prevent the transmission of any pathogenic agents, then one more process may be carried out to remove any degree of discoloration to the water.

Carbon Adsorption - This process removes molecular substances which impart color to a water. This process was described in Chapter 9, and, briefly, takes place in a column filled with granular activated carbon. The carbon surfaces are conditioned to adsorb the colored molecules on the carbon surface. Chapter 9 describes the semi-empirical approach to the design to these columns, scaling up results from small-scale laboratory determinations of adsorption capacity. The addition of this process at the end of the treatment plant will, in addition, filter out any particle breakthough from the previous deep bed filters.

Direct water reuse for drinking water purposes is very rare. South Africa has a full scale plant in operation, Saudi Arabia has a small scale plant in operation, and the City of Denver, Colorado is planning to build and operate such a facility in the next few years.

REFERENCES

[1] Carberry, J. B., and L. R. Stapleford, *J. Amer. Water Works Assoc.*, **71**(4), 213 (1979).

[2] Anon, *Water Quality International*, **3**, 26 (1988).

[3] Montgomery, J. M., Engineers, Inc., *Water Treatment Principles and Design*, John Wiley & Sons, NY, 1985.

[4] Viessman, W., Jr., and C. Welty, *Water Management Technology and Institutions*, Harper & Roe, Publishers, NY, 1985.

[5] Argaman, Y., *J. Amer. Water Works Assoc.*, **63**(12), 775 (1971).
[5] Iwasaki, T., *J. Amer. Water Works Assoc.*, **29**(10), 1591 (1937).

COMPUTER PROBLEMS

Using Equations 10.2 and 10.3 and values of constants from Examples 1 and 2, determine the effects of paddle area and velocity on the required power.

OR

Using Equations 10.4 and 10.5, calculate the headloss and remaining suspended solids concentration values at various bed depths, L, of a 1 meter deep bed sand filter. Use the following values for constants and filter bed characteristics:

$$\mu = 10^{-3}\frac{kg}{m-sec}$$
$$v_{liq} = 5 \text{ m/hr}$$
$$\epsilon = 0.4$$
$$\rho_{liq} = 0.99 \text{ g/cm}^3$$
$$d_m = 0.6 \text{ mm}$$
$$\lambda = 1 \text{ m}^{-1}$$

CONVERSION TABLES

(from Metcalf & Eddy, *Wastewater Engineering, Treatment, and Disposal,*
McGraw Hill, New York, 2nd Edition, 1979)

Metric Conversion Factors (SI units to U.S. customary units)

Multiply the SI unit		by	To obtain the U.S. customary unit	
Name	Symbol		Symbol	Name
Acceleration				
meters per second squared	m/s^2	3.2808	ft/s^2	feet per second squared
meters per second squared	m/s^2	39.3701	in/s^2	inches per second squared
Area				
hectare ($10,000$ m^2)	ha	2.4711	acre	acre
square centimeter	cm^2	0.1550	in^2	square inch
square kilometer	km^2	0.3861	mi^2	square mile
square kilometer	km^2	247.1054	acre	acre
square meter	m^2	10.7639	ft^2	square foot
square meter	m^2	1.1960	yd^2	square yard
Energy				
joule	kJ	0.9478	Btu	British thermal unit
joule	J	2.7778×10^{-7}	$kW \cdot h$	kilowatt-hour
joule	J	0.7376	$ft \cdot lb_f$	foot-pound (force)
joule	J	1.0000	$W \cdot s$	watt-second
joule	J	0.2388	cal	calorie
kilojoule	kJ	2.7778×10^{-4}	$kW \cdot h$	kilowatt-hour
kilojoule	kJ	0.2778	$W \cdot h$	watt-hour
megajoule	MJ	0.3725	$hp \cdot h$	horsepower-hour
Force				
newton	N	0.2248	lb_f	pound force
Flowrate				
cubic meters per day	m^3/d	264.1720	gal/d	gallons per day
cubic meters per day	m^3/d	2.6417×10^{-4}	Mgal/d	million gallons per day
cubic meters per second	m^3/s	35.3147	ft^3/s	cubic feet per second
cubic meters per second	m^3/s	22.8245	Mgal/d	million gallons per day
cubic meters per second	m^3/s	15,850.3	gal/min	gallons per minute
liters per second	L/s	22,824.5	gal/d	gallons per day
liters per second	L/s	0.0228	Mgal/d	million gallons per day
liters per second	L/s	15.8508	gal/min	gallons per minute

Metric Conversion Factors (Continued)

Multiply the SI unit		by	To obtain the U.S. customary unit	
Name	Symbol		Symbol	Name
Length				
centimeter	cm	0.3937	in	inch
kilometer	km	0.6214	mi	mile
meter	m	39.3701	in	inch
meter	m	3.2808	ft	foot
meter	m	1.0936	yd	yard
millimeter	mm	0.03937	in	inch
Mass				
gram	g	0.0353	oz	ounce
gram	g	0.0022	lb	pound
kilogram	kg	2.2046	lb	pound
megagram (10^3kg)	Mg	1.1023	ton	ton (short: 2240 lb)
megagram (10^3kg)	Mg	0.9842	ton	ton (long: 2240 lb)
Power				
kilowatt	kW	0.9478	Btu/s	British thermal units per second
kilowatt	kW	1.3410	hp	horsepower
watt	W	0.7376	ft.lb$_f$	foot-pounds (force) per second
Pressure (force/area)				
pascal (newtons per square meter)	Pa (N/m^2)	1.4504×10^{-4}	lb$_f$/in^2	pounds (force per square inch)
pascal (newtons per square meter)	Pa (N/m^2)	2.0885×10^{-2}	lb$_f$/in^2	pounds(force) per square foot
pascal (newtons per square meter)	Pa (N/m^2)	2.9613×10^{-4}	in Hg	inches of mercury (60^0 F)
pascal (newtons per square meter)	Pa (N/m^2)	4.0187×10^{-3}	in H$_2$O	inches of water (60^0 F)
kilopascal (kilonewtons per square meter)	kPa (kN/m^2)	0.1450	lb$_f$/in^2	pounds (force) per square inch
kilopascal (kilonewtons per square meter)	kPa (kN/m^2)	0.0099	atm	atmosphere (standard)

Metric Conversion Factors (Continued)

Multiply the SI unit		by	To obtain the U.S. customary unit	
Name	Symbol		Symbol	Name
Temperature				
degree Celsius (centigrade)	$^{\circ}$C	$1.8(^{\circ}\text{C}) + 32$	$^{\circ}$F	degree Fahrenheit
degree kelvin	$^{\circ}$K	$1.8(^{\circ}\text{K})-459.67$	$^{\circ}$F	degree Fahrenheit
Velocity				
kilometers per second	km/s	2.2369	mi/h	miles per hour
meters per second	m/s	3.2808	ft/s	feet per second
Volume				
cubic centimeter	cm^3	0.0610	in^3	cubic inch
cubic meter	m^3	35.3147	ft^3	cubic foot
cubic meter	m^3	1.3079	yd^3	cubic yard
cubic meter	m^3	264.1720	gal	gallon
cubic meter	m^3	8.1071×10^{-4}	acre·ft	acre-foot
liter	L	0.2642	gal	gallon
liter	L	0.0353	ft^3	cubic foot
liter	L	33.8150	oz	ounce (U.S. fluid)

Metric Conversion Factors (U.S. customary units to SI units)

Name	Multiply the U.S. customary unit Symbol	by	To obtain the SI unit Symbol	Name
Acceleration				
feet per second squared	ft/s^2	0.3048*	m/s^2	meters per second squared
inches per second squared	in/s^2	0.0254*	m/s^2	meters per second squared
Area				
acre	acre	0.4047	ha	hectare
acre	acre	4.0469×10^{-3}	km^2	square kilometer
square foot	ft^2	9.2903×10^{-2}	m^2	square meter
square inch	in^2	6.4516*	cm^2	square centimeter
square mile	mi^2	2.5900	km^2	square kilometer
square yard	yd^2	0.8361	m^2	square meter
Energy				
British thermal unit	Btu	1.0551	kJ	kilojoule
foot-pound (force)	$ft\cdot lb_f$	1.3558	J	joule
horsepower-hour	$hp\cdot h$	2.6845	MJ	megajoule
kilowatt-hour	$kW\cdot h$	3.600	kJ	kilojoule
kilowatt-hour	$kW\cdot h$	3.600×10^6*	J	joule
watt-hour	$W\cdot h$	3.600*	kJ	kilojoule
watt-second	$W\cdot s$	1.000*	J	joule
Force				
pound force	lb_f	4.4482	N	newton
Flow rate				
cubic feet per second	ft^3/s	2.8317×10^{-2}	m^3/s	cubic meters per second
gallons per day	gal/d	4.3813×10^{-2}	L/s	liters per second
gallons per day	gal/d	3.7854×10^{-2}	m^3/d	cubic meters per day
gallons per minute	gal/min	6.3090×10^{-5}	m^3/s	cubic meters per day
gallons per minute	gal/min	6.3090×10^{-2}	L/s	liters per second
million gallons per day	Mgal/d	43.8126	L/s	liters per second
million gallons per day	Mgal/d	3.7854×10^3	m^3/d	cubic meters per day
million gallons per day	Mgal/d	4.3813×10^{-2}	m^3/s	cubic meters per second

243

Metric Conversion Factors (Continued)

Multiply the SI unit		by	To obtain the U.S. customary unit	
Name	Symbol		Symbol	Name
Length				
foot	ft	0.3048*	m	meter
inch	in	2.54*	cm	centimeter
inch	in	0.0254*	m	meter
inch	in	2.54*	mm	millimeter
mile	mi	1.6093	km	kilimeter
yard	yd	0.9144*	m	meter
Mass				
ounce	oz	28.3495	g	gram
pound	lb	4.5359×10^2	g	gram
pound	lb	0.4536	kg	kilogram
ton (short: 2000 lb)	ton	0.9072	Mg (metric ton)	megagram (10^3 kilogram)
ton (long: 2240 lb)	ton	1.0160	Mg (metric ton)	megagram (10^3 kilogram)
Power				
British thermal units per second	Btu/s	1.0551	kW	kilowatt
foot-pounds (force) per second	ft·lb$_f$/s	1.3558	W	watt
horsepower	hp	0.7457	kW	kilowatt
Pressure (force/area)				
atmosphere (standard)	atm	1.0133×10^2	kPa (kN/m^2)	kilopascal (kilonewtons per square meter)
inches of mercury (60°F)	in Hg (60°F)	3.3768×10^3	Pa (N/m^2)	pascal (newtons per square meter)
inches of water (60°F)	in H$_2$O (60°F)	2.4884×10^2	Pa (N/m^2)	pascal (newtons per square meter)
pounds (force) per square foot	lb$_f$/ft^2	47.8803	Pa (N/m^2)	pascal (newtons per square meter)
pounds (force) per square inch	lb$_f$/in^2	6.8948×10^3	Pa (N/m^2)	pascal (newtons per square meter)
pounds (force) per square inch	lb$_f$/in^2	6.8948	kPa (kN/m^2)	kilopascal (kilonewtons per square meter)
Temperature				
degrees Fahrenheit	°F	$0.555(°F - 32)$	°C	degrees Celsius (centigrade)
degrees Fahrenheit	°F	$0.55(°F + 459.67)$	°K	degree kelvin

244

Metric Conversion Factors (Continued)

Multiply the SI unit		by	To obtain the U.S. customary unit	
Name	Symbol		Symbol	Name
Velocity				
feet per second	ft/s	$0.3048*$	m/s	meters per second
miles per hour	mi/h	$4.4704 \times 10^{-1}*$	m/s	kilometers per second
Volume				
acre-foot	acre-ft	1.2335×10^{3}	m^3	cubic meter
cubic foot	ft^3	28.3168	L	liter
cubic foot	ft^3	2.8317×10^{-2}	m^3	cubic meter
cubic inch	in^3	16.3871	cm^3	cubic centimeter
cubic yard	yd^3	0.7646	m^3	cubic meter
gallon	gal	3.7854×10^{-3}	m^3	cubic meter
gallon	gal	3.7854	L	liter
ounce (U.S. fluid)	oz (U.S. fluid)	2.9573×10^{-2}	L	liter

* Indicates exact conversion.

245

APPENDICES

APPENDIX 1

SATURATION CONCENTRATIONS OF DISSOLVED OXYGEN IN WATER OPEN TO AIR WITH 20.9 % OXYGEN AT A PRESSURE OF 1 ATMOSPHERE

TEMPERATURE	[DO]	Δ [DO] per 100 mg/ℓ Chloride
° C.	mg/ℓ	
0	14.6	0.017
1	14.2	0.016
2	13.8	0.015
3	13.5	0.015
4	13.1	0.014
5	12.8	0.014
6	12.5	0.014
7	12.2	0.013
8	11.9	0.013
9	11.6	0.012
10	11.3	0.012
11	11.1	0.011
12	10.8	0.011
13	10.6	0.011
14	10.4	0.010
15	10.2	0.010
16	10.0	0.010
17	9.7	0.010
18	9.5	0.009
19	9.4	0.009
20	9.2	0.009
21	9.0	0.009
22	8.8	0.008
23	8.7	0.008
24	8.5	0.008
25	8.4	0.008
26	8.2	0.008
27	8.1	0.008
28	7.9	0.008
29	7.8	0.008
30	7.6	0.008

APPENDIX 2

FORMULATION OF THE STREETER-PHELPS EQUATION

CURVE A   Oxygen Depletion $-\dfrac{dy}{dt} = K_1 y = \dfrac{dD_{\text{CurveA}}}{dt}$

where $y = y_0 e^{-K_1 t}$, and

CURVE B   Re-aeration $\dfrac{-dD_{\text{CurveB}}}{dt} = K_2 D$

Total Oxygen Deficit $= \sum \left(\dfrac{dD_{\text{CurveA}}}{dt} + \dfrac{dD_{\text{CurveB}}}{dt} \right)$

and from this summation,

$$\frac{dD}{dt} = K_1 y - K_2 D \tag{1}$$

Rearranging, yields:

$$\frac{dD}{dt} + K_2 D = K_1 y$$

so that the differential is in the classic form:

$$y + Ay = B$$

where, $y' = \dfrac{dD}{dt}$ and $y = D$.

Homogeneous solution gives the following:

$$\frac{dD}{dt} + K_2 D = 0$$

and therefore,

$$\frac{dD}{dt} = -K_2 D$$

and, by separating the variables and integrating,

$$D = Ce^{-K_2 t}$$

Looking for a solution in the following form:

$$D = Ce^{-K_2 t} + Fe^{-K_1 t} \tag{2}$$

so that,

$$\frac{dD}{dt} = -CK_2e^{-K_2t} - FK_1e^{-K_1t} \qquad (3)$$

Substituting the rhs of Equation 1 into the lhs of Equation 3, substituting the rhs of Equation 2 into the resulting term for D in Equation 3, and substituting $y_0e^{-K_1t}$ for y, yields the following expression:

$$(-CK_2e^{-K_2t} - FK_1e^{-K_1t}) + K_2(Ce^{-K_2t} + Fe^{-K_1t}) = K_1y_0e^{-K_1t}$$

Expanding the second term on the lhs yields the following:

$$-CK_2e^{-K_2t} - FK_1e^{-K_1t} + CK_2e^{-K_2t} + FK_2e^{-K_1t} = K_1y_0e^{-K_1t}$$

Therefore,

$$Fe^{-K_1t}(K_1 - K_2) = K_1y_0e^{-K_1t}$$

and,

$$F = \frac{K_1y_0}{K_2 - K_1} \qquad (4)$$

Therefore, by substituting Equation 4 into Equation 2, the following expression results:

$$D = Ce^{-K_2t} + \frac{K_1y_0}{K_2 - K_1}e^{-K_1t} \qquad (5)$$

In order to evaluate C in Equation 5, set $t = 0$, so that $D = D_0$, $D_0 = C + \frac{K_1y_0}{K_2-K_1}$, and $y_0 = y_{\text{mixed}}$ in the case of a stream discharge, where y_{mixed} is the BOD concentration from the discharge immediately after mixing with the more dilute receiving stream.

Therefore, $C = D_0 - \frac{K_1y_{\text{mixed}}}{K_2-K_1}$, and

$$D = \left[D_0 - \frac{K_1y_{\text{mixed}}}{K_2 - K_1}\right]e^{-K_2t} + \frac{K_1y_{\text{mixed}}}{K_2 - K_1}e^{-K_1t}$$

And, finally, by expanding the first term on the rhs and rearranging, the Streeter-Phelps Equation is obtained:

$$D = \frac{K_1y_{\text{mixed}}}{K_2 - K_1}\left[e^{-K_1t} - e^{-K_2t}\right] + D_0e^{-K_2t}$$

249

APPENDIX 3

DIFFERENTIATION OF THE STREETER-PHELPS EQUATION TO OBTAIN t_c

$$D = \frac{K_1 y_{\text{mixed}}}{K_2 - K_1}\left[e^{-K_1 t} - e^{-K_2 t}\right] + D_0\left[e^{-K_2 t}\right] \tag{1}$$

and, differentiating with respect to time yields the following:

$$\frac{dD}{dt} = \frac{K_1 y_{\text{mixed}}}{K_2 - K_1}\left[-K_1 e^{-K_1 t} + K_2 e^{-K_2 t}\right] + D_0\left[-K_2 e^{-K_2 t}\right] \tag{2}$$

Setting the differential equal to zero in order to find the minimum, when $t = t_c$,

$$0 = K_2 e^{-K_2 t_c}\left[\frac{K_1 y_{\text{mixed}}}{K_2 - K_1} - D_0\right] - \frac{K_1^2 y_{\text{mixed}}}{K_2 - K_1}e^{-K_1 t_c}$$

And, rearranging yields the following expression:

$$\frac{K_1^2 y_{\text{mixed}}}{K_2 - K_1}e^{-K_1 t_c} = K_2 e^{-K_2 t_c}\left[\frac{K_1 y_{\text{mixed}}}{K_2 - K_1} - D_0\right]. \tag{3}$$

Dividing both sides of Equation 3 by $e^{-K_2 t_c}$ yields the following expression:

$$\left(\frac{K_1^2 y_{\text{mixed}}}{K_2 - K_1}\right)\left(\frac{e^{-K_1 t_c}}{e^{-K_2 t_c}}\right) = K_2\left[\frac{K_1 y_{\text{mixed}}}{K_2 - K_1} - D_0\right]$$

Simplifying the exponential term yields the following:

$$\frac{K_1^2 y_{\text{mixed}}}{K_2 - K_1}e^{(K_2 - K_1)t_c} = K_2\left[\frac{K_1 y_{\text{mixed}}}{K_2 - K_1} - D_0\right].$$

And transposing the pre-exponential term yields the following:

$$e^{(K_2 - K_1)t_c} = \frac{K_2(K_2 - K_1)}{K_1^2 y_{\text{mixed}}}\left[\frac{K_1 y_{\text{mixed}} - D_0(K_2 - K_1)}{(K_2 - K_1)}\right]$$

Cancelling $K_2 - K_1$ in both numerator and denominator,

$$e^{(K_2 - K_1)t_c} = \frac{K_2}{K_1}\left[\frac{K_1 y_{\text{mixed}} - D_0(K_2 - K_1)}{K_1 y_{\text{mixed}}}\right]$$

And simplifying yields the following:

$$e^{(K_2 - K_1)t_c} = \frac{K_2}{K_1}\left[1 - \frac{D_0(K_2 - K_1)}{K_1 y_{\text{mixed}}}\right].$$

(4)

Taking logs of both sides of Equation 4 yields the following:

$$(K_2 - K_1)t_c = \ln\left\{\frac{K_2}{K_1}\left[1 - \frac{(K_2 - K_1)D_0}{K_1 y_{\text{mixed}}}\right]\right\}$$

and, rearranging yields the final form of the differential:

$$t_c = \frac{1}{K_2 - K_1}\ln\left\{\frac{K_2}{K_1}\left[1 - \frac{D_0(K_2 - K_1)}{K_1 y_{\text{mixed}}}\right]\right\}.$$

APPENDIX 4

MPN INDEX

Number of Tubes Giving Positive Growth out of			MPN Index per 100 ml	Number of Tubes Giving Positive Growth out of			MPN Index per 100 ml
5 of 10 ml each	5 of 1.0 ml each	5 of 0.1 ml each		5 of 10 ml each	5 of 1.0 ml each	5 of 0.1 ml each	
0	0	0	< 2	4	2	1	26
0	0	1	2	4	3	0	27
0	1	0	2	4	3	1	33
0	2	0	4	4	4	0	34
1	0	0	2	5	0	0	23
1	0	1	4	5	0	1	31
1	1	0	4	5	0	2	43
1	1	1	6	5	1	0	33
1	2	0	6	5	1	1	46
1	2	0	6	5	1	2	63
2	0	0	5				
2	0	1	7	5	2	0	49
2	1	0	7	5	2	1	70
2	1	1	9	5	2	2	94
2	2	0	9	5	3	0	79
2	3	0	12	5	3	1	110
				5	3	2	140
3	0	0	8				
3	0	1	11	5	3	3	180
3	1	0	11	5	4	0	130
3	1	1	14	5	4	1	170
3	2	0	14	5	4	2	220
3	2	1	17	5	4	3	280
3	3	0	17	5	4	4	350
4	0	0	13	5	5	0	240
4	0	1	17	5	5	1	350
4	1	0	17	5	5	2	540
4	1	1	21	5	5	3	920
4	1	2	26	5	5	4	1600
4	2	0	22	5	5	5	2400

from *Standard Methods for the Examination of Water and Wastewater*

ALGORITHM FOR THE NUMERICAL INTEGRATION OF TWO LINKED

DIFFERENTIAL EQUATIONS FOR SUBSTRATE AND BIOMASS

(From Forsythe, Malcolm, and Mohler;

Computer Methods For Mathematical Computation,

Prentice Hall, Englewood Cliffs, NJ, 1977.)

```
00100    C--------------------------------------------------------------------------------
00200    C            PROGRAM TO INTEGRATE TWO DIFFERENTIAL EQUATIONS SIMULTANEOUSLY
00300    C            FUNCTIONS FOR SUBSTRATE AND BUGS DEFINED IN LINES 1400:1500
00400    C            ---------------------- FOR M-M BATCH REACTOR  ----------------------
00500            SUBROUTINE ORBIT(T,Y,YP)
00600            REAL T,Y(2),YP(2),R,ALFASQ
00700            REAL THETA,CIN,XIN,ALPHA,XK,YIELD
01100            XK=0.10
01150            KD=0.05
01250            KM=50
01300            YIELD=0.4
01350    C--------------------------------------------------------------------------------
01360    C                    TWO DIFFERENTIAL EQUATIONS IN THE FORM
01370    C                    Y PRIME + AY + B = 0.   FUNCTION PARAMETERS
01380    C                    ARE X AND S LINKED AS Y(2) AND (Y1).
01390    C--------------------------------------------------------------------------------
01400            YP(1)=-XK*Y(1)*Y(2)/(KM+Y(1))
01500            YP(2)=TIELD*Y(1)-KD*Y(2)
01800            RETURN
01900            END
02000    C--------------------------------------------------------------------------------
02100            EXTERNAL ORBIT
02200            REAL T,Y(2),TOUT,RELERR,ABSERR
02300            REAL TFINAL,TPRINT,ECC,ALFA,ALFASQ
02350    C--------------------------------------------------------------------------------
02375    C                    DIMENSION FOR THE NUMBER OF DIFFERENTIAL
02387    C                    EQUATIONS TO BE INTEGRATED SIMULTANEOUSLY,
02393    C            -------------- WORK = 6N + 3, WHERE N = NUMBER OF DE'S  ------------
02395    C--------------------------------------------------------------------------------
02400            REAL WORK (15)
02500            INTEGER IWORK(5),IFLAG,NEQN
02600            COMMON ALFASQ
02700    C*********************************************************************************
02800            OPEN(UNIT=1, DEVICE='DSK', ACCESS='SEQUOT', MODE='ASCII',
02900        *   FILE='CSTR33.DAT')
03000    C*********************************************************************************
03100            NEQN=2
```

```
03200                    T = 0.0
03300                    Y(1) = 250.0
03400                    Y(2) = 2000.0
03500                    RELERR = .0E-08
03600                    ABSERR = 0.0
03700                    TFINAL = 3.0
03800                    TPRINT = .2
03900                    IFLAG = 1
04000                    TOUT = T
04100      010          CALL  RKF45(ORBIT,NEQN,Y,T,TOUT,RELERR,ABSERR,IFLAG,WORK,IWORK)
04200                    WRITE(5,011)  T,Y(1),Y(2)
04300                    WRITE(1,012  T,Y(1),Y(2)
04400                    GO  TO  (080,020,030,040,050,060,070,080),  IFLAG
04500      020          TOUT = T + TPRINT
04600                    IF(T.LT.TFINAL)  GO  TO  010
04700                    STOP
04800      030          WRITE(5,031)  RELERR,ABSERR
04900                    GO  TO  010
05000      040          WRITE(5,041)
05100                    GO  TO  010
05200      050          ABSERR = 1.0E-09
05300                    WRITE(5,031)  RELERR,ABSERR
05400                    GO  TO  010
05500      060          RELERR = 10.0*RELERR
05600                    WRITE(5,031)  RELERR,ABSERR
05700                    IFLAG = 2
05800                    GO  TO  010
05900      070          WRITE(5,071)
06000                    IFLAG = 2
06100                    GO  TO  010
06200      080          WRITE(5,081)
06300                    STOP
06400      C----------------------------------------------------------------------------------------------------------------------
06500      011          FORMAT(F5.1,2E18.4)
06600      012          FORMAT(3E18.4)
06700      031          FORMAT(1H ,'TOLERANCES  RESET',2E12.3)
06800      041          FORMAT(1H ,'MANY  STEPS')
06900      071          FORMAT(1H ,'MUCH  OUTPUT')
07000      081          FORMAT(1H ,'IMPROPER  CALL')
07100                    END
07200                    SUBROUTINE  RKF45(F,NEQN,Y,T,TOUT,RELERR,ABSERR,IFLAG,WORK,IWORK)
07300                    INTEGER  NEQN,IFLAG,IWORK(5)
07400                    REAL  Y(NEQN),T,TOUT,RELERR,ABSERR,WORK(15)
07500                    EXTERNAL  F
07600                    INTEGER  K1,K2,K3,K4,K5,K6,K1M
07700                    K1M = NEQN + 1
```

254

```
07800                K1 = K1M + 1
07900                K2 = K1 + NEQN
08000                K3 = K2 + NEQN
08100                K4 = K3 + NEQN
08200                K5 = K4 + NEQN
08300                K6 = K5 + NEQN
08400                CALL  RKFS(F,NEQN,Y,T,TOUT,RELERR,ABSERR,IFLAG
08500        >       ,WORK(1),WORK(K1M),WORK(K1),WORK(K2),WORK(K3),WORK(K4)
08600        >       ,WORK(K5),WORK(K6),WORK(K6 + 1),IWORK(1),IWORK(2),IWORK(3)
08700        >       ,IWORK(4),IWORK(5))
08800                RETURN
08900                END
09000                SUBROUTINE  RKFS(F,NEQN,Y,T,TOUT,RELERR,ABSERR,IFLAG,YP,H,F1,F2,
09100        *       F3,F4,F5,SAVRE,SAVAE,NFE,KOP,INIT,JFLAG,KFLAG)
09200                LOGICAL  HFAILD,OUTPUT
09300                INTEGER  NEGN,IFLAG,NFE,KOP,INIT,JFLAG,KFLAG
09400                REAL  Y(NEQN),T,TOUT,RELERR,ABSERR,H,YP(NEQN),
09500        *       F1(NEQN),F2(NEQN),F3(NEQN),F4(NEQN),F5(NEQN),SAVRE,
09600        *       SAVAE
09700                EXTERNAL  F
09800                REAL  A,AE,DT,EE,EEOET,ESTTOL,ET,HMIN,REMIN,RER,S,
09900        >       SCALE,TOL,TOLN,U26,EPSP1,EPS,YPK
10000                INTEGER  K,MAXNFE,MFLAG
10100                REAL  AMAX1,AMIN1
10200                DATA  REMIN/1.0E-12/
10300                DATA  MAXNFE/3000/
10400                IF(NEGN.LT.1)  GO  TO  010
10500                IF(RELERR.LT.0.0.OR.ABSERR.LT.0.0.)  GO  TO  010
10600                MFLAG-IABS(IFLAG)
10700                IF(MFLAG.EQ.0.OR.MFLAG.GT.8)  GO  TO  010
10800                IF(MFLAG.NE.1)  GO  TO  020
10900                EPS = 1.0
11000        005     EPS = EPS/2.0
11100                EPSP1 = EPS + 1.0
11200                IF(EPSP1.GT.1.0)  GO  TO  005
11300                U26 = 26.0*EPS
11400                GO  TO  050
11500        010     IFLAG = 8
11600                RETURN
11700        C------------------------------------------------------------------------------------------------------------------------
11800        020     IF(T.EQ.TOUT,AND.KFLAG.NE.3)  GO  TO  010
11900                IF(MFLAG.NE.2)  GO  TO  025
12000                IF(KFLAG.EQ.3.OR.INIT.EQ.0)  GO  TO  045
12100                IF(KFLAG.EQ.4)  GO  TO  040
12200                IF  (KFLAG.EQ.5.AND.ABSERR.EQ.0.0)  GO  TO  030
12300                IF(KFLAG.EQ.6.AND.RELERR.LE.SAVRE.AQND.ABSERR.LE.SA
```

255

```
12400                 >      VAE)  GO  TO  030
12500                        GO  TO  050
12600          025          IF(IFLAG.EQ.3)  GO  TO  045
12700                        IF(IFLAG.EQ.4)  GO  TO  040
12800                        IF(IFLAG.EQ.5.AND.ABSERR.GT.0.0)  GO  TO  045
12900          030          STOP
13000          040          NFE=0
13100                        IF(MFLAG.EQ.2)  GO  TO  050
13200          045          IFLAG=JFLAG
13300                        IF(KFLAG.EQ.3)  MFLAG=TABS(IFLAG)
13400          050          JFLAG=IFLAG
13500                        KFLAG=0
13600                        SAVRE=RELERR
13700                        SAVAE=ABSERR
13800                        RER=2.0*EPS+REMIN
13900                        IF(RELERR,GE.RER)  GO  TO  055
14000                        RELERR=RER
14100                        IFLAG=3
14200                        KFLAG=3
14300                        RETURN
14400          C-----------------------------------------------------------------------------------------------
14500          055          DT=TOUT-T
14600                        IF(MFLAG.EQ.1)  GO  TO  060
14700                        IF(INIT.EQ.0)  GO  TO  065
14800                        GO  TO  080
14900          060          INIT=0
15000                        KOP=0
15100                        A=T
15200                        CALL  F(A,Y,YP)
15300                        NFE=1
15400                        IF(T.NE.TOUT)  GO  TO  065
15500                        IFLAG=2
15600                        RETURN
15700          C-----------------------------------------------------------------------------------------------
15800          065          INIT=1
15900                        H=ABS(DT)
16000                        TOLN=0.0
16100                        DO  070  K-1,NEQN
16200                        TOL=RELERR*ABS(Y(K))+ABSERR
16300                        IF(TOL.LE.0.0)  GO  TO  070
16400                        TOLN=TOL
16500                        YPK=ABS(YP(K))
16600                        IF(YPK*H**5.GT.TOL)  H=(TOL/YPK)**0.2
16700          070          CONTINUE
16800                        IF(TOLN.LE.0.0)  H=0.0
16900                        H=AMAX1(H,U26*AMAX1(ABS(T),ABS(DT)))
```

```
17000                   JFLAG = ISIGN(2,IFLAG)
17100        080        H = SIGN(H,DT)
17200                   IF(ABS(H).GE.2.0*ABS(DT))  KOP = KOP + 1
17300                   IF(KOP.NE.100)  GO  TO  085
17400                   KOP = 0
17500                   IFLAG = 7
17600                   RETURN
17700        C--------------------------------------------------------------------------------
17800        085        IF(ABS(DT).GT.U26*ABS(T))  GO  TO  095
17900                   DO  090  K = 1,NEQN
18000        090        Y(K) = Y(K) + DT*YP(K)
18100                   A = TOUT
18200                   CALL  F(A,Y,YP)
18300                   NFE = NFE + 1
18400                   GO  TO  300
18500        095        OUTPUT = .FALSE.
18600                   SCALE = 2.0/RELERR
18700                   AE = SCALE*ABSERR
18800        100        HFAILD = .FALSE.
18900                   HMIN = U26*ABS(T)
19000                   DT = TOUT-T
19100                   IF(ABS(DT).GE.2.0*ABS(H))  GO  TO  200
19200                   IF(ABS(DT).GT.ABS(H))  GO  TO  150
19300                   OUTPUT = .TRUE.
19400                   H = DT
19500                   GO  TO  200
19600        150        H = 0.5*DT
19700        200        IF(NFE.LE.MAXNFE)  GO  TO  220
19800                   IFLAG = 4
19900                   KFLAG = 4
20000                   RETURN
20100        C--------------------------------------------------------------------------------
20200        220        CALL  FEHL(F,NEQN,Y,T,H,YP,F1,F2,F3,F4,F5,F1)
20300                   NFE = NFE + 5
20400                   EEOET = 0.0
20500                   DO  250  K = 1,NEQN
20600                   ET = ABS(Y(K)) + ABS(F1(K)) + AE
20700                   IF(ET.GT.0.0)  GO  TO  240
20800                   IFLAG = 5
20900                   RETURN
21000        C--------------------------------------------------------------------------------
21100        240        EE = ABS((-2090.0*YP(K) + (21970.0*F3(K)-15048.0*F4(K))) +
21200             >     (Q22528.0*F2(K)-27360.0*F5(K)))
21300        250        EEOET = AMAX1(EEOET,EE/ET)
21400                   ESTTOL = ABS(H)*EEOET*SCALE/752400.0
21500                   IF(ESTTOL.LE.1.0)  GO  TO  260
```

257

```
21600                    HFAILD = .TRUE.
21700                    OUTPUT = .FALSE.
21800                    S = 0.1
21900                    IF(ESTTOL.LT.59049.0)  S = 0.9/ESTTOL**0.2
22000                    H = S*H
22200                    IFLAG = 6
22300                    KFLAG = 6
22400                    RETURN
22500       C----------------------------------------------------------------------------------
22600       260          T = T + H
22700                    DO  270  K = 1,NEQN
22800       270          Y(K) = F1(K)
22900                    A = T
23000                    CALL  F(A,Y,YP)
23100                    NFE = NFE + 1
23200                    S = 5.0
23300                    IF(ESTTOL.GT.1.889568E-04)  S = 0.9/ESTTOL**0.2
23400                    IF(HFAILD)  S = AMIN1(S,1.0)
23500                    H = SIGN(AMAX1(S*ABS(H),HMIN),H)
23600       C************************************************************************************
23700                    IF(OUTPUT)  GO  TO  300
23800                    IF(IFLAG.GT.0)  GO  TO  100
23900                    IFLAG = -2
24000                    RETURN
24100       C----------------------------------------------------------------------------------
24200       300          T = TOUT
24300                    IFLAG = 2
24400                    RETURN
24500       C----------------------------------------------------------------------------------
24600                    END
24700       CXXXXXXXXXXXXXXXXXXXXXXXXXXXXXXXXXXXXXXXXXXXXXXXXXXXXXXXXXXXXXXXX
```

```
24800                SUBROUTINE  (FEHL(F,NEQN,Y,T,H,YP,F1,F2,F3,F4,F5,S)
24900                INTEGER  NEQN
25000                REAL  Y(NEQN),T,H,YP(NEQN,F1(NEGN),F2(NEQN),
25100           >    F3(NEQN),F4(NEQN),F5(NEQN),S(NEQN)
25200                REAL  CH
25300                INTEGER  K
25400                CH = H/4.0
25500                DO  221  K = 1,NEQN
25600      221       F5(K) = Y(K) + CH*YP(K)
25700                CALL  F(T + CH,F5,F1)
25800                CH = 3.0*H/32.0
25900                DO  222  K = 1,NEQN
26000      222       F5(K) = Y(K) + CH*(YP(K) + 3.0*F1(K))
26100                CALL  F(T + 3.0*H/8.0,F5,F2
26200                CH = H/2197.0
26300                DO  223  K = 1,NEQN
26400      223       F5(K) = Y(K) + CH*(1932.0*YP(K) + (7296.0*F2(K)-7200.0*F1(K)))
26500                CALL  F(T + 12.0*H/13.0,F5,F3)
26600                CH = H/4104.0
26700                DO  224  K = 1,NEQN
26800      224       F5(K) = Y(K) + CH*((8341.0*YP(K)-845.0*F3(K)) +
26900           >    (29440.0*F2(K)-32832.0*F1(K)))
27000                CALL  F(T + H,F5,F4)
27100      C = = = = = = = = = = = = = = = = = = = = = = = = = = = = = = = = = = = = = = = = = = = = = = = = = =
27200                CH = H/20520.0
27300                DO  225  K = 1,NEQN
27400      225       F1(K) = Y(K) + CH*((-6080.0*YP(K) + (9295.0*F3(K)-
27500           >    5643.0*F4(K))) + (41040.0*F1(K)-28352.0*F2(K)))
27600                CALL  F(T + h/2.0,F1,F5)
27700      C = = = = = = = = = = = = = = = = = = = = = = = = = = = = = = = = = = = = = = = = = = = = = = = = = =
27800                CH = H/7618050.0
27900                DO  230  K = 1,NEQN
28000      230       S(K)Y(K) + CH*((902880.0*YP(K) + (3855735.0*F3(K)-
28100           >    1371249.0*F4(K))) + (3953664.0*F2(K) +
28200           >    277020.0*F5(K)))
28300                RETURN
28400      C-------------------------------------------------------------------------------
28500                END
28600      CXXXXXXXXXXXXXXXXXXXXXXXXXXXXXXXXXXXXXXXXXXXXXXXXXXXXXXXXXXXXXXXX
```

INDEX

complex order, 118.
Confirmed Test, 105, 107.
continuous flow, 140, 189, 191.

Death rate constant, 113.
detention time, 145, 151–153, 159, 160.
dewatering, 187, 197–201.
digestion, 187, 194–197, 202.
disinfection, 245, 249, 250.
dissociation, 67–72, 74, 75, 77, 78, 100.
dissolved oxygen, 37, 53, 55.
DO depletion, 53, 66.
Drinking Water Standards, 47, 65, 232, 233.
drying beds, 197.
drying, 187, 197, 200, 202.

Electrolyte, 98, 99, 209, 216.
electroneutrality, 74, 75.
electrophoretic mobility, 93.
endogenous, 194, 195.
energy barrier, 94, 96, 97.
enumeration, 107, 110.
epilimnion, 25, 26.
equilibrium, 2, 3, 18, 33, 115, 116.
equivalence point, 71, 76, 78, 88.
eutrophication, 30, 31, 34.
exerted BOD, 54–56.

Filtration, 240, 245, 246, 248.
first order reaction, 20, 49, 117, 119–121, 125, 128, 130–134, 139, 141, 187, 246, 248.
floc, 95, 99, 101, 179, 183, 190, 197, 216, 242, 244, 245.

Grit chamber, 144, 145.
growth rate, 113.

Hardness, 87, 88, 89.
head loss, 245–248.
heavy metals, 46.
heterogeneous, 9, 67, 128, 133, 161.
heterotrophic, 11–17, 26, 28, 33.
homogeneous, 67.
hydration layer, 93, 94.
hydraulic detention time, 133, 136.
hydraulic load, 162, 163, 165, 166, 169–171, 173, 185, 186.
hydrogen ion, 68–72, 74–77, 99.
hydrogen sulfide, 44.
hydrologic cycle, 6, 33.
hydrophilic, 90, 91, 100.
hydrophobic, 91, 93, 94, 98, 100.
hydroxyl ion, 70, 71, 74, 75, 77, 87.
hypolimnion, 25, 26, 28.

Incineration, 200, 201, 204.
indicator organism, 46, 49, 56, 104, 107, 114.
inflection point, 71, 75, 192.
interface, 190.

Kinetics, 116, 122, 128, 131, 138.

Lag, 54, 115.
lagoon, 162–164, 202, 213.
lime, 76–78, 87, 89, 166, 170, 203, 209, 216, 243.
limiting concentration, 189.
limiting solids flux, 192.
linear flow velocity, 145.

Mass balance, 129, 130,

stable, 93–97, 99, 216, 221.
steady state, 3, 33.
stoichiometric coefficients, 116,
    117.
stratification, 26, 28, 34.
subsidence rate, 189.
substrate uptake rate, 113.
substrate, 11, 12, 16, 28.
sulfate, 44, 46, 58.
sulfur, 35, 44.
surface of shear, 94.
surface properties, 67, 89, 90.
suspension, 9, 90, 93–101, 146,
    148, 150–152, 154–156,
    160, 179, 216, 219,
    245–247, 249.

Temperature profile, 25, 26,
    34.
tertiary treatment, 142, 143.
thermocline, 25, 26.
thermodynamic, 115.
thickening, 187, 189, 190,
    192–194, 206.
titration, 58, 71, 72, 74–78, 82,
    88.
total solids flux, 190.
trickling filter, 163–166,
    168–173, 185, 212.
turbidity, 37, 49.

Ultimate BOD, 53, 65.

Vacuum filtration, 197, 198.
velocity gradient, 242–244.
volumetric flow rate, 21, 23,
    137, 151, 159, 168, 183,
    184.

Water Quality Index, 37.

zero order reaction, 117, 118,
    125, 130, 132.

zeta potential, 93, 94, 99.